SMART SOLUTIONS

Skills, Problem Solving, Tools, and Applications

Comprehensive Math Review

Cathy Fillmore Hoyt

New Readers Press

Acknowledgments

Advisers to the Series

Connie Eichhorn
Supervisor of Transitional Services
Omaha Public Schools
Omaha, NE

Mary B. Puleo
Assistant Director
Sarasota County Adult and
Community Education
Sarasota, FL

Lois Kasper
Instructional Facilitator
N.Y. Board of Education
New York, NY

Margaret Rogers
Coordinator
San Juan Unified Adult Education
Sacramento, CA

Jan Phillips
Assistant Professor
William Rainey Harper College
Palatine, IL

Library of Congress Cataloging-in-Publication Data

Hoyt, Cathy Fillmore, date.
Comprehensive math review / Cathy Fillmore Hoyt.
p. cm. — (Smart solutions)
ISBN 1-56420-126-0 (pbk.)
1. Mathematics. I. Title. II. Series.
QA39.2.H69   1997
513'.14'076—dc20   96-36810
                                                        CIP

ISBN 1-56420-126-0

Copyright © 1997
New Readers Press
U.S. Publishing Division of Laubach Literacy International
Box 131, Syracuse, New York 13210-0131

Printed in the United States of America

Director of Acquisitions and Development: Christina Jagger
Photo Illustrations: Mary McConnell
Developer: Learning Unlimited, Oak Park, IL
Developmental Editor: Kathy Osmus
Copy Editor: Judi Lauber
Design: Katie Bates, Patricia A. Rapple

9 8 7 6 5 4 3 2

Contents

Introduction

Math skills play an increasingly vital role in today's world. Everyone needs to work confidently with numbers to solve problems on the job and in other areas of daily life.

This book and the others in the *Smart Solutions* series can help you meet your everyday math needs. Each unit is organized around four key areas that will build your competence and self-confidence:

- **Skills** pages present instruction and practice with both computation and word problems.
- **Tools** pages provide insight into how to use objects (such as rulers or calculators) or apply ideas (such as estimates or equations) to a wide variety of math situations.
- **Problem Solver** pages provide key strategies to help you become a successful problem solver.
- **Application** pages are real-life topics that require mathematics.

Key Features

Skill Preview: You can use the Skill Preview to determine what skills you already have and what you need to concentrate on.

Talk About It: At the beginning of each unit, you will have a topic to discuss with classmates. Talking about mathematics is key to building your understanding.

Key Concepts: Throughout the book, you will see this symbol ▶, which indicates key math concepts and rules.

Making Connections: Throughout each unit, you will work with topics that connect math ideas to various interest areas and to other math concepts.

Special Problems: These specially labeled problems require an in-depth exploration of math ideas. You may be asked to explain or draw or to do something else that demonstrates your math skills.

Working Together: At the end of each unit, you will work with a partner or small group to apply your math skills.

Mixed and Unit Reviews: Periodic checkups will help you see how well you understand the material and can apply what you have learned.

Posttest: At the end of the book, you will find a test that combines all of the book's topics. You can use this final review to judge how well you have mastered the book's skills and strategies.

Glossary: Use this list of terms to learn or review key math words and ideas.

Tool Kit: You can refer to these resource pages as you work through the book.

Skill Preview

This survey of math skills will help you and your teacher decide what you need to study to get the most out of this book. It will show you how much you already know and what you need to learn.

Do as much as you can of each section below. If you can't do all of the problems in a section, go ahead to the next section and do all of the problems that you can.

Part 1: Whole Numbers

Solve the following problems.

1. $\begin{array}{r} 7,605 \\ + 4,986 \end{array}$

2. 714×8

3. $\begin{array}{r} 3,004 \\ - 1,839 \end{array}$

4. $14,392 \div 7$

5. $\begin{array}{r} 8,605 \\ \times\ 52 \end{array}$

6. $25\overline{)4,555}$

7. Nasser and Farah are driving from Denver to Kansas City, a distance of 606 miles. They drive 334 miles to Hays, Kansas, on the first day of the trip. How much farther do they have left to drive?

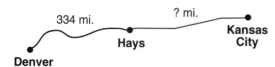

334 mi.   ? mi.

Denver   **Hays**   **Kansas City**

8. Educe has two job offers. The first job pays $27,000 per year. The second pays $2,375 per month. Which job pays more? Explain your answer.

9. Ticket prices for an amusement park are shown on the sign below. How much will a family of 5 (2 adults and 3 children) pay for admission to the park?

Magic Land Park

Admission   Adults $23.00
Children $16.00

10. Yuki kept a log of the hours she worked for the last 4 weeks. Find the total hours she worked during that period.

Week 1	45
Week 2	42
Week 3	38
Week 4	46

11. Alex puts $125 per month in a savings account. At that rate, how much will he put into the account over 18 months?

12. A department store recorded 25,916 sales in a month. If there were 31 days in the month, how many sales did the store average per day?

Part 2: Decimals and Money

Write each of the following values using a decimal point.

13. nine hundredths

15. forty dollars and five cents

14. two dollars and ninety-eight cents

16. thirty-five thousandths

Solve the following problems.

17. $100 - 84.7$

19. $8.3 + 9.56 + 10.8$

21. 6.3×8

18. 54×0.006

20. $36 \div 0.08$

22. $1.8 \div 5$

23. A wheel assembly is made from 2 wheels, each weighing 2.6 pounds, and a steel axle, which weighs 1.8 pounds. What is the total weight of the assembly?

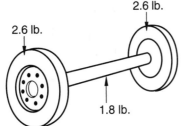

2.6 lb.

2.6 lb.

1.8 lb.

24. Nick owes $42.78 for a new car antenna. He hands the cashier three $20 bills. How much does he receive in change?

25. Tanya saw the sign below at a music store. She bought 3 CDs and 4 tapes at the prices shown. How much was her purchase before sales tax?

GOING OUT OF BUSINESS

All CDs $7.89

All Tapes $4.69

26. If an airplane travels 2.5 hours at an average speed of 630 miles per hour, how many total miles will it travel?

Part 3: Fractions, Ratios, and Percents

Choose the number with the larger value in each pair.

27. $\frac{1}{8}$ $\frac{9}{100}$

29. 0.09   0.085

31. 1%   0.1

28. $1\frac{1}{4}$ $\frac{12}{8}$

30. 0.25   0.3

32. $\frac{3}{2}$ 125%

Solve the following problems. Simplify fractions where possible.

33. $\frac{3}{5} + \frac{1}{4} + \frac{7}{10}$

35. $1\frac{3}{4} \div \frac{3}{8}$

37. $60 \times \frac{3}{4}$

34. $\frac{7}{8} - \frac{1}{2}$

36. $\frac{2}{3} \times \frac{3}{4}$

38. $40 \div \frac{1}{2}$

Solve the following problems.

39. A ton of ore is found to contain 20% copper. At that rate, how many tons of ore will be needed to get 10 tons of copper?

40. A & B Retail gives a discount for purchases over certain amounts.

For purchases of at least	$150	$275	$400
Discount rate	6%	12%	18%

Using the chart above, find the dollar amount of the discount given to a customer whose purchase totals $312.

41. Nita spent $340 for materials to make a sculpture. She sold the sculpture for 250% of the cost of the materials. What was the selling price of the sculpture?

42. Risa and Marco are packing shipping crates. After 30 minutes, Risa has packed 5 crates and Marco has packed 4. Working together, how long will it take them to pack 108 crates?

43. Adnan mixes 2 parts peat moss with 5 parts potting soil to plant rose bushes. How much peat moss should he mix with 35 cubic feet of potting soil?

Peat moss   Potting soil

44. Brett, a football player, made 18 field goals out of 22 attempted field goals. To the nearest whole percent, what percent of the attempted field goals were successful?

Part 4: Data and Measurement

Change each quantity to the unit(s) indicated.

45. 54 in. = _____ ft. _____ in.

46. $3\frac{2}{3}$ yd. = _____ ft.

47. 165 cm = _____ m

48. 32 fl. oz. = _____ c.

49. $3\frac{1}{2}$ gal. = _____ qt.

50. 2,000 g = _____ kg

Solve each problem.

51. Which metric unit would most likely be used to measure these items:

   a. gasoline: _____

   b. the medicine in a pill: _____

   c. distance between cities: _____

52. LaYell has two bags of candy, weighing 50 ounces and 70 ounces. How many pounds of candy does she have?

Use the chart to answer problem 53.

Exercise	Calories Burned per Hour
Basketball	750
Running, 12-min. mile	650
Walking, 5 mi./hr.	555
Tennis, singles	425

53. Jim plays basketball on Monday for 2 hours and walks (at a rate of 5 miles per hour) for 1 hour per day Tuesday through Friday. How many calories does he burn off from these forms of exercise per week?

Use the graph below to solve problems 54 and 55.

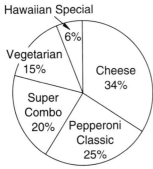

Percent of Pizzas Ordered by Type

Pizza Bob offers five kinds of pizza. The circle graph at left shows the percent of each type of pizza ordered.

54. If 1,500 pizzas were ordered last month, how many cheese pizzas were ordered?

55. What percent of the orders are made up of the 3 most popular pizzas?

Part 5: Algebra and Geometry

Find the value of each of the following expressions.

56. $8^2 =$

57. $\frac{-75}{15} =$

58. $5(-6 + 3) =$

Solve the following equations and inequalities.

59. $7(y - 5) = 21$

60. $11 - 2n = 5$

61. $-4(2x - 3) = -36$

62. $3(y - 6) > 27$

63. $3x + 5(x - 2) < 3(x - 5)$

64. $8c \geq 2c + 18$

Solve the following problems.

65. Find the value of $3x + 8$ when $x = 5$.

66. Solve for A in the expression $A = lw$ where $l = 18$ and $w = 24$.

67. Write and simplify an expression for the perimeter of the rectangle shown here.

68. Soo Ji needs to score at least 360 points on 4 tests to earn an A in her math class. On the first 3 tests she scores 92, 86, and 88. What is the least she can score on the final test to earn an A?

Problems 69 and 70 refer to the following figure.

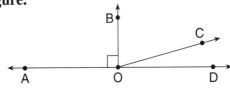

69. Name a supplementary angle to $\angle AOB$.

70. $\angle BOC$ measures $75°$. What is the measure of $\angle COD$?

71. A tree casts a shadow of 12 feet. At the same time, a 3-foot stake, which is perpendicular to the ground, casts a 4-foot shadow. Find the height of the tree.

1. 12,591

2. 5,712

3. 1,165

4. 2,056

5. 447,460

6. 182 R5

7. **272 miles**

   $606 - 334 = 272$

8. **The second pays more.**

   First job per month: $27,000 ÷ 12 = $2,250

   Since $2,375 is greater than $2,250, the second job pays more.

   or

   Second job per year: $2,375 × 12 = $28,500

   Since $28,500 is greater than $27,000, the second job pays more.

9. **$94**

   $2($23) + 3($16) = $46 + $48 = 94

10. **171 hours**

    $45 + 42 + 38 + 46 = 171$

11. **$2,250**

    $$125 \times 18 = $2,250$

12. **836 sales**

    $25,916 ÷ 31 = 836$

13. 0.09

14. $2.98

15. $40.05

16. 0.035

17. 15.3

18. 0.324

19. 28.66

20. 450

21. 50.4

22. 0.36

23. **7 pounds**

    $2(2.6) + 1.8 = 5.2 + 1.8 = 7.0$

24. **$17.22**

    $3($20) - $42.78 = $60 - $42.78 = 17.22

25. **$42.43**

    $3($7.89) + 4($4.69) = $23.67 + $18.76 = 42.43

26. **1,575 miles**

    $630 \times 2.5 = 1,575$

27. $\frac{1}{8}$

28. $\frac{12}{8}$

    $\frac{12}{8} = 1\frac{1}{2}$

29. **0.09**

    0.090 is greater than 0.085.

30. **0.3**

    0.30 is greater than 0.25.

31. **0.1**

    0.1 = 0.10 = 10%, which is greater than 1%, or 0.01.

32. $\frac{3}{2}$

    $\frac{3}{2} = 1\frac{1}{2} = 150\%$, which is greater than 125%.

33. $1\frac{11}{20}$

$\frac{3}{5} + \frac{1}{4} + \frac{7}{10} = \frac{12}{20} + \frac{5}{20} + \frac{14}{20} = \frac{31}{20} = 1\frac{11}{20}$

34. $\frac{3}{8}$

$\frac{7}{8} - \frac{1}{2} = \frac{7}{8} - \frac{4}{8} = \frac{3}{8}$

35. $4\frac{2}{3}$

$1\frac{3}{4} \div \frac{3}{8} = \frac{7}{4} \div \frac{3}{8} = \frac{7}{4} \times \frac{8}{3} = \frac{14}{3} = 4\frac{2}{3}$

36. $\frac{1}{2}$

37. 45

$\frac{60}{1} \times \frac{3}{4} = 45$

38. 80

$40 \div \frac{1}{2} = 40 \times \frac{2}{1} = 80$

39. **50 tons of ore**

One way to solve this problem is to set up a proportion.

$\frac{\text{amount of ore}}{\text{copper yield}} \quad \frac{1 \text{ ton of ore}}{0.2 \text{ ton of copper}} = \frac{x \text{ tons of ore}}{10 \text{ tons of copper}}$

$\frac{1}{.2} = \frac{x}{10}$

$\frac{10}{.2} = x$

$50 = x$

Another approach is to use a percent statement:

20% of $x = 10$ tons

$.2x = 10$

$x = \frac{10}{.2}$

$x = 50$

You can also think 20% $= \frac{1}{5}$, so out of every ton of ore, $\frac{1}{5}$ is copper. Therefore, it must take 5 tons of ore to make 1 ton of copper. To make 10 tons of copper, you need 5×10, or 50 tons of ore.

40. **$37.44**

The customer qualifies for a 12% discount.

12% of $312 = \$312 \times 0.12 = \$37.44$

41. **$850**

250% of $340 = \$340 \times 2.50 = \$850$

42. **360 minutes or 6 hours**

Together they can pack 9 boxes in 30 minutes or $\frac{1}{2}$ hour (.5 hour). Set up a proportion and solve.

$\frac{9 \text{ boxes}}{.5 \text{ hour}} = \frac{108 \text{ boxes}}{x \text{ hours}}$

$\frac{108(.5)}{9} = 6 = x$

43. **14 cubic feet of peat moss**

$\frac{2 \text{ parts peat moss}}{5 \text{ parts potting soil}} = \frac{x}{35}$

$5x = 2(35)$

$\frac{2(35)}{5} = \frac{70}{5} = 14$

44. **82%**

$18 \div 22 \approx 0.818 \approx 0.82 = 82\%$

45. **4 ft. 6 in.**

$54 \div 12 = 4 \text{ R6}$

46. **11 ft.**

$3\frac{2}{3} \times 3 = 11$

47. **1.65 m**

$165 \div 100 = 1.65$

48. **4 c.**

$32 \div 8 = 4$

49. **14 qt.**

$3\frac{1}{2} \times 4 = 14$

50. **2 kg**

$2,000 \div 1,000 = 2$

51. a. liters

   b. milligrams

   c. kilometers

52. $7\frac{1}{2}$ **lb.**

$50 + 70 = 120 \div 16 = 7.5 \text{ or } 7\frac{1}{2}$

53. **3,720 calories**

$2(750) + 4(555) = 1,500 + 2,220 = 3,720$

54. 510 cheese pizzas

34% of $1,500 = 1,500 \times 0.34 = 510$

55. 79%

$34\% + 25\% + 20\% = 79\%$

56. $8 \cdot 8 = $ **64**

57. **–5**

58. **–15**

$5(-6 + 3) = 5(-3) = -15$

59. $y = $ **8**

$7(y - 5) = 21$

$7y - 35 = 21$

$7y = 56$

$y = 8$

60. $n = $ **3**

$11 - 2n = 5$

$-2n = -6$

$n = \frac{-6}{-2}$

$n = 3$

61. $x = $ **6**

$-4(2x - 3) = -36$

$-8x + 12 = -36$

$-8x = -48$

$x = 6$

62. $y > $ **15**

$3(y - 6) > 27$

$3y - 18 > 27$

$3y > 45$

$y > 15$

63. $x < $ **–1**

$3x + 5(x - 2) < 3(x - 5)$

$3x + 5x - 10 < 3x - 15$

$8x - 10 < 3x - 15$

$5x < -5$

$x < -1$

64. $c \geq $ **3**

$8c \geq 2c + 18$

$6c \geq 18$

$c \geq 3$

65. 23

$3x + 8 = 3(5) + 8 = 15 + 8 = 23$

66. $A = $ **432**

$A = lw$

$A = 18(24) = 432$

67. $6x$

$x + 2x + x + 2x = 6x$

68. 94 points

$92 + 86 + 88 + x \geq 360$

$266 + x \geq 360$

$x \geq 94$

69. $\angle$BOD or $\angle$DOB

70. 15°

$90 - 75 = 15$

71. 9 feet

$\frac{3 \text{ ft. stake}}{4 \text{ ft. shadow}} = \frac{x \text{ ft. tree}}{12 \text{ ft. shadow}}$

$4x = 3(12)$

$\frac{3(12)}{4} = \frac{36}{4} = 9$

Skill Preview Diagnostic Chart

Make note of any problems that you answered incorrectly. Notice the skill area for each problem you missed. As you work through this book, be sure to focus on these skill areas.

Problem Number	Skill Area	Unit
1, 2, 3, 4, 5, 6	Whole number operations	1
7, 8, 9, 10, 11, 12	Word problems with whole numbers	1
13, 14, 15, 16	Writing decimals and monetary amounts	2
17, 18, 19, 20, 21, 22	Decimal operations	2
23, 24, 25, 26	Word problems with decimals	2
27, 28, 29, 30, 31, 32	Comparing fractions, decimals, and percents	3
33, 34, 35, 36, 37, 38	Fraction operations	3
40, 41, 44	Solving percent problems	3
39, 42, 43	Solving ratio and proportion problems	3
45, 46, 48, 49, 52	Using the English measurement system	4
47, 50, 51	Using the metric measurement system	4
53	Reading tables	4
54, 55	Reading graphs	4
56, 57, 58, 65, 66, 67	Evaluating expressions	5
59, 60, 61	Solving equations	5
62, 63, 64, 68	Solving inequalities	5
69, 70	Angles	5
71	Similar figures	5

Whole Number Review

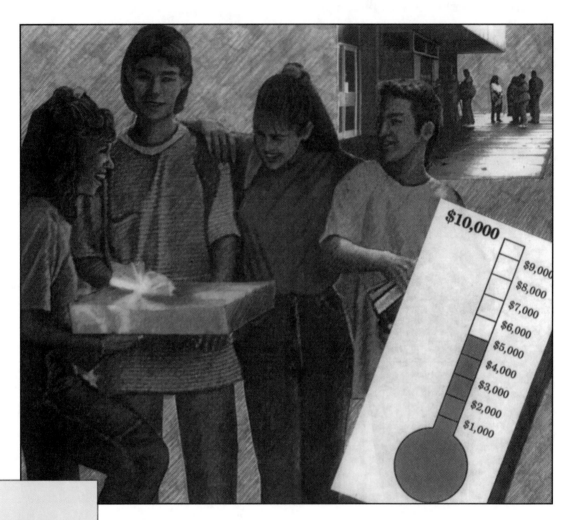

Whole numbers are the numbers we use for counting. As you know, our counting system uses 10 digits: 0, 1, 2, 3, 4, 5, 6, 7, 8, and 9. Any number, no matter how large, can be written with these 10 digits.

The value of a digit depends on its place in a number. **Place value** is the basis for our system of reading and writing numbers and making calculations.

In this unit, you'll apply whole numbers in many practical situations. You'll use your knowledge of place value to estimate and solve for answers.

When Do I Use Whole Numbers?

Each of the following situations uses whole numbers. Check the experiences you have had.

- ☐ figuring the miles between two cities
- ☐ keeping track of pounds lost or gained
- ☐ saving money for a large purchase
- ☐ counting up the hours you've worked in a week
- ☐ figuring the number of people coming to a party
- ☐ estimating each person's share of a group gift

If you checked any of the situations described above, you have used whole numbers. Read the following items, and describe some of your experiences on the lines provided.

1. Think about the last time you went to the grocery store. How much money did you spend? Was the total amount closer to $10 or $100? Explain how you know.

   _____

   _____

2. Explain how you could figure out how long it would take you to drive a certain number of miles. Have you ever done this?

   _____

   _____

3. Have you ever split the cost of a meal with a group of friends? How did you figure out each person's share?

   _____

   _____

4. Describe how you could use house numbers or street numbers to figure out how much farther you have to go to reach a certain address.

   _____

   _____

Talk About It

Think of a project you have completed recently. How did you budget your time? How did you estimate how long the project would take? How did you decide how much you could get done each day? Discuss your methods with other students in your class.

Addition

You add to combine numbers to find a total amount. If the **sum** of a place value is 10 or more, you need to **regroup.** Write the last digit under the column you are adding and regroup (carry) the extra digit to the column to the left. Before you add numbers, you should **estimate** to get an idea of what the answer should be. Use your estimate to see if your answer makes sense.

Adding and Regrouping

Example: Sasha had $479. She earned $28 more. How much money does she have now?

Estimate first.
Round to the nearest 100.
└──► 500
+ 30 ◄──┐
530   └─ Round to the nearest 10.

Step 1
Add the ones. The ones add up to 17 ones. Write 7 in the ones column. Regroup 1 ten to the tens place.

$$\begin{array}{r} \overset{1}{4}79 \\ +\ 28 \\ \hline 7 \end{array}$$

Step 2
Add the tens. The tens add up to 10 tens. Write 0 in the tens column. Regroup the 1 hundred to the hundreds place.

$$\begin{array}{r} \overset{11}{4}79 \\ +\ 28 \\ \hline 07 \end{array}$$

Step 3
Add the hundreds.

$$\begin{array}{r} \overset{11}{4}79 \\ +\ 28 \\ \hline 507 \end{array}$$

Check: The answer **$507** is close to your estimate of $530.

A. **Add, regrouping if necessary. Estimate first.**
Note: ≈ means "is approximately equal to."

Estimate

1. $\begin{array}{r} 58 \\ +\ 21 \end{array}$ ≈ $\begin{array}{r} 60 \\ +\ 20 \end{array}$ $\begin{array}{r} 63 \\ +\ 15 \end{array}$ $\begin{array}{r} \$32 \\ +\ 55 \end{array}$ $\begin{array}{r} \$12 \\ +\ 37 \end{array}$

2. $302 + $95 $473 + $125   109 + 650   211 + 120 + 38

$\begin{array}{r} \$302 \\ +\ 95 \end{array}$ ◄── Line up ones, tens, and hundreds.

3. $\begin{array}{r} \$3,435 \\ +\ 1,250 \end{array}$ $\begin{array}{r} 9,750 \\ +\ 148 \end{array}$ $\begin{array}{r} 74 \\ +\ 18 \end{array}$ $\begin{array}{r} \$19 \\ 35 \\ +\ 50 \end{array}$

4. 137 + 845   358 + 803 $635 + $98   348 + 75 + 709

5. $1,085   6,532 $5,938 $3,299
   + 3,927   + 7,085   + 546   + 985

B. Solve the problems below. Estimate an answer first.

6. Lorna took her son to the doctor. The visit cost her $48. She also paid $28 for a prescription. How much did Lorna spend?

Estimate: $50 + $30 = _____

Exact: $48 + $28 = _____

7. Local schools and businesses worked together to clean up a five-mile stretch of the coastline. The schools gathered 2,379 pounds of garbage. The businesses cleaned up 4,096 pounds. What was the total number of pounds of garbage collected?

8. In his first three full seasons in the major leagues, Juan Gonzalez hit 27, 43, and 46 home runs. What was his home run total for the three years?

9. On Friday, Craig and Nita drove from Charlotte to Birmingham, a distance of 391 miles. The next day they drove from Birmingham to Jackson, a 245-mile drive. How many total miles did they drive on Friday and Saturday?

10. Two recent business closures put many people in California out of work. An aircraft company laid off 569 employees. When a food company closed, 392 people lost their jobs. How many jobs were lost because of these closures?

Use this chart to solve problems 11 and 12.

Fall Carnival Fundraising Event	
Booth	**Money Raised**
Fish Pond.	$185
Skeeball	649
Basketball Shoot	298
Bean Bag Toss	368
Miniature Golf	572
Dunk Tank	935

11. What was the total amount of money raised by the two most profitable booths at the Fall Carnival?

12. Estimate The chairperson for the event predicted that the Dunk Tank would earn more than the Fish Pond, Basketball Shoot, and Bean Bag Toss combined. Use estimation to decide whether the chairperson's prediction came true.

Subtraction

Subtraction is used to find the **difference** between numbers. To subtract, put the larger number on top. If a digit in the top number is smaller than the digit below it, you need to **regroup** (borrow) from a higher place value column.

Subtracting and Regrouping

Example: The Software Barn has 3,025 software titles in stock. The store sold 287 software titles on Monday. How many titles are still in stock?

Estimate first.	Step 1	Step 2	Step 3
	Subtract the ones. Since $5 < 7^*$, regroup 1 ten. $10 + 5 = 15$ $15 - 7 = 8$	Subtract the tens. Since $1 < 8$, regroup. First regroup to the hundreds column, then to the tens column.	Continue subtracting, working from right to left.

$$\begin{array}{r} 3{,}000 \\ -\ 300 \\ \hline 2{,}700 \end{array} \qquad \begin{array}{r} 3{,}02\overset{1}{5} \\ -\ 287 \\ \hline 8 \end{array} \qquad \begin{array}{r} 3{,}025 \\ -\ 287 \\ \hline 38 \end{array} \qquad \begin{array}{r} 3{,}025 \\ -\ 287 \\ \hline 2{,}738 \end{array}$$

Check: The answer **2,738** is close to your estimate of 2,700.

$^*5 < 7$ means "5 is less than 7." The symbol < means "is less than."

$7 > 5$ means "7 is greater than 5." The symbol > means "is greater than."

$79 \approx 80$ means "79 is approximately equal to 80." The symbol ≈ means "is approximately equal to."

A. Subtract. Regroup if necessary. Estimate first.

1.
$$\begin{array}{r} \text{Estimate} \\ 79 \approx \ 80 \\ -\ 52 \approx -\ 50 \\ \hline \end{array} \qquad \begin{array}{r} 38 \\ -\ 16 \\ \hline \end{array} \qquad \begin{array}{r} \$146 \\ -\ 24 \\ \hline \end{array} \qquad \begin{array}{r} 547 \\ -\ 305 \\ \hline \end{array}$$

2. $895 - \$75$ $3{,}298 - 1{,}285$ $\$6{,}175 - \$5{,}025$ $9{,}982 - 850$

$$\begin{array}{r} \$895 \\ -\ 75 \\ \hline \end{array}$$

3.
$$\begin{array}{r} 36 \\ -\ 18 \\ \hline \end{array} \qquad \begin{array}{r} 72 \\ -\ 39 \\ \hline \end{array} \qquad \begin{array}{r} \$850 \\ -\ 66 \\ \hline \end{array} \qquad \begin{array}{r} \$294 \\ -\ 178 \\ \hline \end{array}$$

4. 507 – 395 $900 – $775 $400 – $89   1,723 – 784

5. $2,050   5,000 $7,200 $3,035
       –   975   – 1,220   – 5,610   – 1,850

B. Solve the problems below. Estimate an answer first.

6. In 1993, Matt Williams had 43 home runs before a strike cut the baseball season short. How many more home runs did he need to tie the major league record of 61 home runs in one season?

Estimate: 60 – 40 = _____

Exact: 61 – 43 = _____

7. Freida's take-home pay is $824 per month. How much will she have left after she pays her rent of $385?

8. Carl is saving money to buy a stereo system. The system he wants is $605. He has saved $380. How much more does he need to buy the system?

9. At its deepest point, Lake Superior is 1,330 feet deep. The depth of Lake Michigan is 407 feet less. How deep is Lake Michigan at its deepest point?

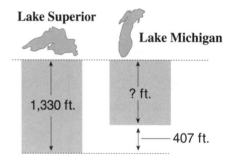

C. In a newspaper story on basketball, you find this chart. Answer these questions using the information on the chart.

10. Michael Jordan was the NBA's leading scorer for seven years. Find the difference between his highest and lowest point totals.

11. Wilt Chamberlain was the NBA's leading scorer in 1962 with 4,029 points. No other basketball player has scored that many points in one year. Find the difference in points between Chamberlain's record and Jordan's best year.

12. **Discuss** In which years did Jordan's point total decline from the year before? Discuss with a partner different reasons that could account for the lower totals in those years.

Michael Jordan
Chicago Bulls
NBA Scoring Leader

Year	Points
1987	3,041
1988	2,868
1989	2,633
1990	2,753
1991	2,580
1992	2,404
1993	2,541

Multiplication

Like addition, **multiplication** often requires regrouping. When a product of two digits is 10 or more, regroup to the next place value column. Multiply by the digit in the next place, then *add* the regrouped amount.

Multiplying and Regrouping

Example: Find the cost of 9 months of health insurance at $428 a month.

Estimate first.	Step 1	Step 2	Step 3
	Multiply the ones. The product is 72 ones. Regroup 7 to the tens place.	Multiply the tens. Add the regrouped 7. $2 \times 9 = 18$, and $18 + 7 = 25$ Regroup 2 to the hundreds place.	Multiply the hundreds. Add the regrouped 2. $4 \times 9 = 36$, and $36 + 2 = 38$

$$\begin{array}{r} 400 \\ \times\ 10 \\ \hline 4{,}000 \end{array}$$

$$\begin{array}{r} {}^{7}\ \\ \$42\overset{7}{8} \\ \times\ \ 9 \\ \hline 2 \end{array}$$

$$\begin{array}{r} {}^{2\,7} \\ \$428 \\ \times\ \ 9 \\ \hline 52 \end{array}$$

$$\begin{array}{r} {}^{2\,7} \\ \$428 \\ \times\ \ 9 \\ \hline \$3{,}852 \end{array}$$

Check: The answer **$3,852** is close to your estimate of 4,000.

A. Estimate, then find an exact answer. Regroup when necessary.

1.
Estimate				
$34 \approx 30$	58	165	392	2,085
$\underline{\times 2} \approx \underline{\times 2}$	$\underline{\times 4}$	$\underline{\times 5}$	$\underline{\times 8}$	$\underline{\times\ \ 7}$

When multiplying by a two-digit number, multiply by the ones digit to get a **partial product.** Then multiply by the tens digit to get a second partial product. Begin the second partial product in the tens place. (You can use a placeholder zero in the ones place.) Add the two partial products.

Multiplying by Larger Numbers

Example: A company pays a $25 rebate to 182 customers. How much did it pay out in rebates? Estimate: $200 \times \$20 = \$4,000$

Step 1	Step 2	Step 2
Multiply by the ones digit.	Multiply by the tens digit.	Add the partial products.

larger number on top ⟶
$$\begin{array}{r} {}^{4\,1} \\ 182 \\ \times\ 25 \\ \hline 910 \end{array}$$

$$\begin{array}{r} {}^{1} \\ 182 \\ \times\ 25 \\ \hline 910 \\ 3640 \end{array}$$ ⟵ placeholder 0

$$\begin{array}{r} 182 \\ \times\ 25 \\ \hline 910 \\ {}^{1}\ \\ 3\,640 \\ \hline \$4{,}550 \end{array}$$

Check: The answer **$4,550** is close to the estimate of $4,000.

B. Estimate first. Then multiply the numbers below.

Estimate

2. 68 ≈ 70 $352   409   525   819
   × 16 ≈ × 20   × 9   × 27   × 5   × 38

3. $83 × 50   572 × 42   1,035 × 75   7,809 × 45   82 × 13,521

4.   75 $1,035 $250   20,947   8,068
   × 75   ×   4   × 26   ×   12   × 42

C. Solve the problems below. Estimate an answer first.

5. The Cougar fan club reserved a section of seats at an away game. The section included 22 rows of 38 seats each. How many seats are in the section?

   Estimate: 20 × 40 = _____

   Exact: 22 × 38 = _____

6. Rocio has $12 withheld from each weekly paycheck for health insurance. How much will be withheld in a year for health insurance? (One year is 52 weeks.)

7. A school is storing water for an emergency. The school officials order 4 dozen bottles of water. If each bottle contains 5 gallons, how many gallons of water does the school have on hand?

8. Jim is driving across Arizona on Highway 60. Because of high winds, he averages only 35 miles per hour. At that rate, how many miles can he drive in 4 hours?

D. Use the graph to solve the problems.

The owner of Cal's Diner made a graph to show the average number of customers served each week during each shift.

9. How many customers will the diner likely serve in 4 weeks during the breakfast shift?

10. According to the graph, about how many more customers will the diner serve during the lunch shift than the supper shift over the course of a year (52 weeks)?

11. **Investigate** The owner of Cal's Diner needs to cut back from 3 shifts to 2 shifts. The owner estimates that the diner makes $3 per customer for breakfast, $2 per customer for lunch, and $2 per customer for supper. Which shift makes the smallest amount of money and should therefore be cut?

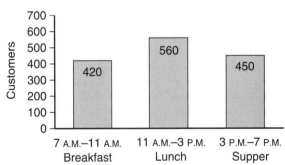

Cal's Diner
Weekly Customers per Shift

Division

Long division is a four-step process that repeats until there are no digits left to bring down. Any nonzero number left over after the final step is called a **remainder.** The remainder must always be less than the number you divided by.

▶ **Four-Step Division Process**

Step 1. Divide and write the answer as part of the quotient.

Step 2. Multiply the answer by the divisor.

Step 3. Subtract.

Step 4. Bring down the next digit and repeat the process if needed.

Using the Four-Step Process

Example: Frank needs to pack 426 computer chips in packages of 8. How many whole packages can he make? Estimate: 400 ÷ 8 = 50

Step 1	**Step 2**	**Step 3**	**Step 4**
Divide 8 into 42. Write the answer above the 2.	Multiply: 5 × 8 = 40	Subtract: 42 − 40 = 2	Bring down the next digit. Repeat.

Step 1:
$$8\overline{)426} \quad \text{with } 5 \text{ above}$$

Step 2:
$$\begin{array}{r} 5 \\ 8\overline{)426} \\ 40 \end{array}$$

Step 3:
$$\begin{array}{r} 5 \\ 8\overline{)426} \\ 40 \\ \hline 2 \end{array}$$

Step 4:
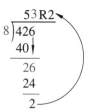
$$\begin{array}{r} 53\,R2 \\ 8\overline{)426} \\ 40 \\ \hline 26 \\ 24 \\ \hline 2 \end{array}$$

He can make **53** whole packages. He will have 2 chips left over.

Check: Your answer of 53 is reasonably close to 50.

The remainder is 2. Write the remainder after the letter *R*.

A. **Estimate first and then find an exact answer. Be sure to include remainders (if any) in your answers.**

Estimate

1. $5\overline{)135}$ $5\overline{)150}$ $9\overline{)1,790}$ $8\overline{)1,640}$ $4\overline{)3,520}$

2. 8,890 ÷ 3   2,435 ÷ 2   6,064 ÷ 4   13,962 ÷ 7

$3\overline{)8,890}$

3. $4\overline{)1,517}$ $5\overline{)8,941}$ $6\overline{)3,090}$ $9\overline{)10,420}$

4. 3,175 ÷ 2   7,040 ÷ 5   18,306 ÷ 9   310,500 ÷ 8

B. Solve the problems below. Estimate an answer first.

5. A hotel offered a company a special rate of $702 for 9 rooms for 1 night. How much did the company pay per room?

   Estimate: $700 ÷ 10 = _____

   Exact: $702 ÷ 9 = _____

6. Isaac bought a carton of 1,200 candy bars to resell. If he packages them in boxes of 8, how many boxes will he have to sell?

7. Besides their regular wages, 3 sales clerks split a percentage of the total sales during a store's Midnight Madness sale. If the percentage they split was $1,095, how much more did each clerk earn?

8. An employer estimates a job will take 128 hours. The job needs to be finished in 1 workday. She plans to hire temporary employees to work 8 hours each. How many temps will she need to do the job?

Making Connections: Using Compatible Numbers

What's the best way to estimate an answer to a division problem?

The attendance at a movie theater for a 4-day run of a movie was 2,584. On average, how many people attended the movie each day of the run?

You need to divide 2,584 by 4 to solve the problem. You could round 2,584 to 3,000 and divide by 4 (3,000 ÷ 4 = 750). A simpler way to estimate might be to use compatible numbers instead of rounding to the nearest hundred or thousand. Two numbers are **compatible** if one is a multiple of the other.

You know that 4 × 6 is 24. 2,584 ≈ 2,400.
You can do this division in your head.

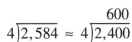

$$4\overline{)2,584} \approx 4\overline{)2,400}\ \ ^{600}$$

Now compare the results.

Rounding	Compatible Numbers	Exact Answer
3,000 ÷ 4 = 750	2,400 ÷ 4 = 600	2,584 ÷ 4 = 646

Compatible numbers are easy to use and provide reasonable estimates.

Use compatible numbers to estimate. Then solve for the exact answer.

1. 14,986 ÷ 5

   Estimate: _____

   Exact: _____

2. 6,530 ÷ 8

   Estimate: _____

   Exact: _____

3. 47,600 ÷ 9

   Estimate: _____

   Exact: _____

4. 47,600 ÷ 7

   Estimate: _____

   Exact: _____

Division by Two or More Digits

When you divide by two or more digits, you use both educated guessing and estimation to help you find the answer.

Dividing by a Large Number

Example: Mike needs to save $4,316 in 1 year (52 weeks). How much will he need to save each week? Estimate: $4,500 ÷ 50 = $90

Step 1
52 won't go into 4 or 43. Estimate how many times 52 goes into 431. Using basic division facts, look for compatible numbers. Using lead digits from 52 and 431, think, "How many 5s are in 40?" Try 8 times 52.

$$\begin{array}{r} 8 \\ 52\overline{)4,316} \end{array}$$

Step 2
Multiply: 52 × 8 = 416
Subtract. Bring down the 6. Using lead digits from 52 and 156, think, "How many 5s are in 15?" Try 3 times 52.

$$\begin{array}{r} 83 \\ 52\overline{)4,316} \\ \underline{4\,16} \\ 156 \end{array}$$

Step 3
Multiply: 52 × 3 = 156
Subtract.

$$\begin{array}{r} 83 \\ 52\overline{)4,316} \\ \underline{4\,16} \\ 156 \\ \underline{156} \\ 0 \end{array}$$

Check: Your answer of **$83** is close to the estimate of $90.

A. Divide.

1. $42\overline{)84}$ $36\overline{)540}$ $19\overline{)152}$ $23\overline{)322}$ $93\overline{)3,384}$

2. $52\overline{)700}$ $26\overline{)1,000}$ $18\overline{)2,250}$ $55\overline{)13,695}$ $13\overline{)11,100}$

Sometimes a zero is needed in the answer to keep the correct place value. You can use an estimate to check place value.

Putting Zeros in Answers

Example: Lei has 1,630 toy figures to display. She can fit 8 toys on a metal rack. How many racks will she need? Estimate: 1,600 ÷ 8 = 200

Step 1
16 ÷ 8 = 2

$$\begin{array}{r} 2 \\ 8\overline{)1,630} \\ \underline{16} \\ 0 \end{array}$$

Step 2
Bring down the 3. Since 8 won't go into 3, put a 0 in the answer above the 3.

$$\begin{array}{r} 20 \\ 8\overline{)1,630} \\ \underline{16} \\ 03 \end{array}$$

Step 3
Bring down the 0 and divide 30 by 8.

$$\begin{array}{r} 203 \text{ R6} \\ 8\overline{)1,630} \\ \underline{16} \\ 030 \\ \underline{24} \\ 6 \end{array}$$

Step 4
Decide what to do with the remainder. Lei will completely fill 203 racks. She will need 1 more rack to hold the remaining 6 toys.
203 + 1 = 204

Lei needs **204** metal racks, which is close to your estimate of 200.

B. Divide. Watch out for zeros.

3. $12\overline{)2,460}$ $\qquad$ $9\overline{)20,736}$ $\qquad$ $26\overline{)22,360}$ $\qquad$ $30\overline{)7,625}$

4. $15\overline{)9,078}$ $\qquad$ $25\overline{)201,025}$ $\qquad$ $38\overline{)3,040}$ $\qquad$ $140\overline{)84,280}$

C. Solve the problems below. Estimate an answer first.

5. Mikail earned $1,632 for 4 weeks of work. He worked 160 hours during the month. How much did Mikail earn per week?

Estimate: $1,600 ÷ 4 = _____

Exact: $1,632 ÷ 4 = _____

6. Renée needs to cut 20 equal lengths of ribbon from a roll containing 2,100 centimeters. What will be the length of each of the 20 ribbons?

7. Joan scored 252 points during a basketball season. She played in 18 games. On average, how many points did she score per game?

8. A total of 220,460 cars crossed the Lincoln Toll Bridge over the course of a year. What is the average number of cars that crossed the bridge in a day? (*Hint:* There are 365 days in a year.)

D. Use the graph to answer the questions below.

**Marisol and Edwin Ruiz
Annual Expenses
Total Income — $41,300**

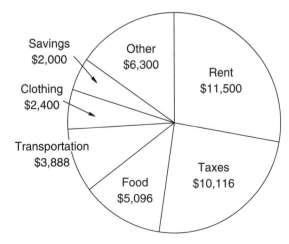

Savings $2,000
Other $6,300
Rent $11,500
Clothing $2,400
Transportation $3,888
Food $5,096
Taxes $10,116

9. Taxes are withheld from the couple's paychecks at an even rate throughout the year. How much do they pay in taxes in 1 month?

10. Next year the Ruizes would like to save twice as much, although they expect their income will be the same. They decide to cut back on food, transportation, clothing, and other expenses to save money. If they cut the same dollar amount on each of the 4, how much will they spend on clothing next year?

11. Explain The amount the Ruizes spend on rent is nearly triple the expense of one other item. Which item is it? Explain how you found your answer.

Mental Math and Estimation

Tools

$x - 7 = 3$

Estimation is a very useful math tool. As you have seen, a good estimate is a useful way to check an answer. In some situations, a good estimate is all the answer that you will need.

Estimating is most helpful when you can do all the work in your head. The goal in estimation is to work with "friendly" numbers. Usually, you will use rounding to make the work easier. But remember, there is no single right way to estimate. In the examples below, three different ways of estimating are shown.

Rounding and Estimating

Example 1: $3,206 \times 9$

Use front-end estimation. Round to the first digit on the left.

Remember: To multiply numbers that end in 0, multiply the leading digits ($3 \times 1 = 3$). Then add on the number of zeros in the problem.

$$\begin{array}{rcll} 3,206 & \approx & 3,000 & \text{3 zeros} \\ \times \quad 9 & \approx & \times \quad 10 & \text{1 zero} \\ \hline 28,854 & & 30,000 & \text{4 zeros} \end{array}$$

Example 2: $14,340 \div 6$

Use compatible numbers. Watch for relationships between numbers.

$$6\overline{)14,340}^{\,2,390} \approx 7\overline{)14,000}^{\,2,000}$$

or

$$6\overline{)14,340}^{\,2,390} \approx 6\overline{)12,000}^{\,2,000}$$

Example 3: $26,379 + 24,278$

Use your knowledge of money. If you use nickels, dimes, and quarters, you're used to working with multiples of 5, 10, and 25. Both the numbers in this problem are close to 25,000. Because you know $25 + 25 = 50$, you know the answer will be close to 50,000.

$$\begin{array}{rcl} 26,379 & \approx & 25,000 \\ + \, 24,278 & \approx & + \, 25,000 \\ \hline 50,657 & & 50,000 \end{array}$$

Check: In each case, the estimate was close to the exact answer.

A. **Estimate an answer using any method. Then choose the exact answer from the given choices. Finally, solve for the exact answer to see if your estimate was helpful.**

Estimate

1. $\begin{array}{rcl} 492 & \approx & 500 \\ \times \, 19 & \approx & \times \, 20 \end{array}$

2. $\begin{array}{r} 5,149 \\ 4,895 \\ + \, 5,068 \end{array}$ Estimate

3. $55\overline{)1,210}$ Estimate

(1) 3,688

(2) 9,348

(3) 13,118

(1) 14,092

(2) 15,112

(3) 16,232

(1) 6

(2) 14

(3) 22

26

B. Use your estimation skills to solve the following problems.

4. Rheena earns $1,912 per month. About what does she earn per year?

   (1) between $18,000 and $20,000

   (2) between $20,000 and $22,000

   (3) between $22,000 and $24,000

5. Marie and her 3 brothers split the cost of sending their parents to Hawaii for their 40th anniversary. The cost of the trip was $3,575. Which of the following amounts is closest to each child's share?

   (1) $800

   (2) $900

   (3) $1,000

6. Central City plans to acquire new buses to improve its public transportation system. It currently has 15,890 regular riders. The city wants to buy 1 new bus for every 2,000 regular riders. About how many buses will the system need?

   (1) 5 or 6 buses

   (2) 7 or 8 buses

   (3) 9 or 10 buses

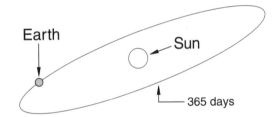

Making Connections: When Is an Estimate Enough?

The Earth is 93 million miles from the Sun and takes 365 days to orbit the Sun. Which of these numbers is rounded? They both are. The distance from the Sun is rounded to the millions place, and the time needed for the Earth's orbit is rounded to the nearest day. Both are rounded, but the second number is more accurate.

Exact answers are needed in order to make exact calculations. Rounded numbers are enough to make comparisons and communicate ideas.

Use a ✔ to show whether you would want a rounded or an exact number for each situation. Discuss your choices with a partner.

Rounded	Exact

1. The number of people injured in an industrial accident.
2. The number of Americans who own bicycles.
3. The average cost of a home in Southern California.
4. The score of a football game.
5. The record-high temperature for the summer.
6. The number of people who attended a parade.

The Five-Step Plan

You have already been doing word problems in this book. A **word problem** is a story with numbers that asks a question. To solve a word problem, you need to get a "sense" of the problem.

Instead of this:

$$\begin{array}{r} 325 \\ -\ 210 \end{array}$$

You have this:

On Monday, Fast Copy had 325 cases of white paper in stock. On Saturday, it had 210 cases in stock. The company plans to order 200 more cases next week. How many cases of white paper did Fast Copy use during the week?

To succeed with word problems, follow the five-step plan below.

Step 1. Understand the question.
What are you being asked to find? Put the question in your own words.

If Fast Copy started with 325 cases of paper and had 210 cases left at the end of the week, how many cases were used?

Step 2. Decide what information is needed to solve the problem.
Word problems may give you more information than you need or not enough information.

Needed: 325 cases in stock on Monday
          210 cases left on Saturday
Not needed: 200 cases, the number to be
          ordered next week.

Step 3. Think about how you might solve the problem.
- To combine, add.
- To find a difference, subtract.
- To find a product, multiply.
- To find a part, divide.

$$\begin{array}{r} 325 \\ -\ 210 \end{array}$$ cases in stock on Monday
cases left on Saturday
cases used

Step 4. Estimate an answer.
About how large or small should the answer be?
Use approximate numbers that are easy to work with.

$325 \approx 300$
$210 \approx 200$
$300 - 200 = 100$
$\approx$ means "is approximately equal to"

Step 5. Solve the problem. Check your answer.
Do the computation. Then look at your answer.
Does it make sense? Is it a reasonable solution to the problem?

$$\begin{array}{r} 325 \\ -\ 210 \\ \hline \mathbf{115} \end{array}$$ cases on hand
cases left
cases used

Check: 115 is close to the estimate, and it is a reasonable solution.

Use the five-step problem-solving plan to answer the following questions.
Follow each step as shown.

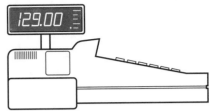

1. Stuart has $310 in his budget to buy computer supplies for his job. He pays $99 for a publishing program at a software store. At an office supply store, he buys a printer cartridge and diskettes for $129. How much does he have left in his budget?

Step 1. Understand the question.
What are you being asked to find?

   a. After spending $99 and $129, how much does Stuart have left?

Step 2. Decide what information is needed to solve the problem.
What numbers do you need?

   b.

Step 3. Think about how you might solve the problem.
Choose the operation. Set up the problem.

   c.

Step 4. Estimate an answer.
Round the numbers.

   d.

Step 5. Solve. Check your answer.
Does the answer make sense?

   e.

2. Carrie bought 2 dozen chocolate donuts. She also bought 8 donuts with powdered sugar and 10 glazed donuts. How many donuts did she buy in all?

   a. What is the question?
   b. What information is needed?
   c. What strategy should you use?
   d. Estimate an answer.
   e. What is the solution? Is it sensible?

3. Debra and Kate share an apartment. They share the monthly rent of $386 evenly. Last month their phone bill was $42. How much does each woman spend per month on rent?

   a. What is the question?
   b. What information is needed?
   c. What strategy should you use?
   d. Estimate an answer.
   e. What is the solution? Is it sensible?

4. Tim is building a closet. One sheet of plywood costs $26. Tim will need 4 sheets of plywood to do the job. He will also need to buy stain and varnish at $12 per gallon. How much will he spend on plywood?

   a. What is the question?
   b. What information is needed?
   c. What strategy should you use?
   d. Estimate an answer.
   e. What is the solution? Is it sensible?

Using Your Calculator

A **calculator** can make solving problems easier. Most calculators are similar to the one shown here. The keys may be in different places, but that's OK. All calculators have the same **digit keys** and the four **function keys.** Using these keys, you'll be able to do all the exercises in this book.

In some exercises, you will be asked to use a calculator. You should also feel free to use your calculator throughout this book to check answers and to work problems with many numbers or hard numbers.

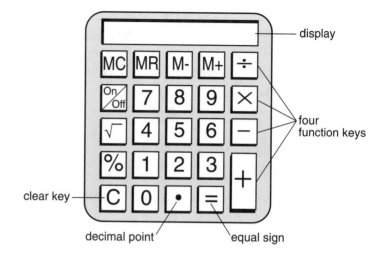

The Calculator Display

Calculators have a digital display that shows values in whole numbers and decimals. Most displays can show up to 8 digits at once. When a whole number is entered, the decimal point is shown at the end of the number.

When you turn on your calculator, the digital display reads 0. Each digit you enter will also appear on the display.

Entering Digits on a Calculator

Example 1: How do you enter the number 75?

Your display reads:

Step 1. Turn on the calculator. Press the $\boxed{7}$ key. Then press the $\boxed{5}$ key.

| 75. |

Step 2. Press $\boxed{C}$. The clear key "clears" the display.

| 0. |

Example 2: Enter the number 1,020.

Step 1. Press $\boxed{1}\boxed{0}\boxed{2}\boxed{0}$. You don't need to enter the comma.

| 1020. |

Step 2. Press $\boxed{C}$ to clear the display.

| 0. |

Now try some basic computing with your calculator.

Computing with Two Numbers

Example: What is 236 plus 95?

Press these keys in order:	**Your display reads:**
2 **3** **6**	236.
+	236.
9 **5**	95.
=	331.

The answer shown on your display should be **331.**

Computing with More than Two Numbers

Example: What is 350 and 89 subtracted from 1,000?

Press these keys in order:	**Your display reads:**
1 **0** **0** **0**	1000.
−	1000.
3 **5** **0**	350.
−	650.
8 **9**	89.
=	561.

The answer shown on your display should be **561.**

 Estimate an answer to each problem. Then use a calculator to find the exact answer.

1. 904 + 68   13 × 8   1,204 ÷ 86   3,500 − 499

2. 3,100 ÷ 62   53,290 − 9,087   375 × 15   98 + 16 + 105

3. 820 − 589 − 85   6 × 9 × 15   8,899 ÷ 11   29 + 84 + 8 + 36

4. Marta needs to till, seed, and fertilize 1,500 square yards of ground. On Monday, she finishes 390 square yards. On Tuesday, she completes 425 square yards. How many more square yards are left to do?

5. **Investigate** Last year Paul earned $15,000 as a data entry clerk. This year, after receiving a raise, he earns $1,375 per month. How much more will Paul earn this year than last year? Find two ways to solve the problem.

Unit 1 Review

A. Estimate. Then solve each problem.

1.
$$45 + 32$$
$$481 + 316$$
$$\$1{,}056 + 725$$
$$3{,}957 + 2{,}505$$
$$\$12{,}945 \\ 5{,}470 \\ + 10{,}550$$

2.
$$97 - 63$$
$$\$189 - 56$$
$$352 - 28$$
$$1{,}920 - 375$$
$$\$4{,}000 - 2{,}199$$

3.
$$43 \times 2$$
$$265 \times 3$$
$$\$1{,}096 \times 5$$
$$89 \times 26$$
$$\$6{,}575 \times 18$$

4. $9\overline{)1{,}152}$ $5\overline{)4{,}560}$ $6\overline{)650}$ $12\overline{)7{,}428}$ $25\overline{)70{,}230}$

B. Estimate the answer to each problem. Then use your calculator to find the exact answer.

5. $18{,}600 \times 32$ $7{,}200 - 660$ $2{,}884 \div 14$

   Estimate: _____

   Exact: _____

6. $16{,}254 \div 9$ $69 + 75 + 183$ 504×97

7. $\$1{,}798 + \$3{,}098$ $72{,}040 \div 8$ $39{,}460 - 15{,}595$

C. Choose the correct method to solve each problem.

8. Alex borrowed $5,000 to buy a used car. He agreed to make monthly payments of $176 for 36 months. How much will he actually pay the bank?

(1) $5,000 − $176 − 36

(2) $5,000 ÷ 36

(3) $176 × 36

9. Sandra earns $7 per hour as a part-time receptionist. For word processing, she earns $13 per hour. How much will she earn working 25 hours at a word processing job?

(1) $13 × 25

(2) $13 + $7 × 25

(3) $13 + $7 + 25

10. Midvale School has 575 students enrolled. Last Friday 492 students were present. How many students were absent?

(1) 575 + 492

(2) 575 − 492

(3) 575 ÷ 492

11. When Nita picked up her rental car, the odometer showed 65,847. Nita drove 169 miles the first day. At the end of the second day, the odometer showed 66,411. How many miles did Nita drive the second day?

(1) 66,411 − 65,847

(2) 66,411 − 65,847 − 169

(3) 66,411 − 65,847 + 169

 D. Use the information on the chart to solve the problems. You may use a calculator.

Book World, Fall Inventory October 14

Section	Books in Stock
Romance	847
Science Fiction	1,224
Adventure	69
Non-Fiction	1,197
Reference	1,872
Self Help	752
Literature	1,685

12. On September 1, Book World had 2,100 books in stock for each section. What was the total number of books in stock?

13. On October 14, the manager placed orders so that each section would contain 2,400 books. How many books did the manager order for the Literature section?

14. Evaluate If no book shipments arrived between September 1 and October 14, which type of book was most popular during that time? Which was least popular? Explain.

Working Together

Estimate the number of hours of television you watch each day. Add to find your total for the week. Working in a small group, share your statistics. To find the average weekly total for the group, add the weekly totals for the members of the group and divide by the number of people in the group. Discuss whether you think this average would be representative of the whole class; of other age groups.

Unit 2

Decimals and Money

Skills

Understanding and writing decimals

Comparing decimals

Adding decimals

Subtracting decimals

Multiplying decimals

Dividing decimals

Tools

Calculators

Problem Solvers

Solving multistep problems

Choosing a method

Applications

Relating decimals and money

Unit price and total cost

The **decimal system** represents whole and decimal amounts. When a digit is written to the right of a decimal point, that digit has a value that is *less than 1*.

In the U.S. money system, parts of a dollar are shown using decimals. The amount $5.75 means 5 *whole* dollars and 75 *hundredths* of a dollar.

In this unit, you'll apply decimals in many practical situations. You'll work with money and measurements. You will also see how using a calculator can make your work easier.

When Do I Use Decimals?

Each of the following activities uses decimals. Check the experiences you have had.

- ☐ paying bills
- ☐ purchasing gasoline
- ☐ balancing a checkbook
- ☐ making change
- ☐ reading baseball batting averages
- ☐ reading metric measurements on packages

If you've done any of those things, you've used decimals. Describe some of your experiences on the lines below.

1. Describe a recent bill you paid. Did you write a check? Write the total amount paid, including cents. (Cents are a form of decimals.)

   _____

   _____

2. Have you ever used a digital scale to weigh food at the grocery store? What would a weight of 3.5 pounds mean? Explain how you know.

   _____

   _____

3. Have you ever punched a time clock? The payroll office keeps track of your hours using decimals. What does it mean if your time card shows you have worked 7.75 hours? Explain how you know.

   _____

   _____

Talk About It

Each time you make a purchase and receive change, you are working with decimals. How do you make sure you are getting the right change? What are some of the different methods used to make change? Discuss your ideas with others.

Throughout this unit, you may use a calculator wherever that would be useful.

Understanding Decimals

Each time you use money, you are working with decimals. In the U.S. money system, a **decimal point** separates dollars from cents. Numbers after the decimal point represent a value *less than $1*.

Similarly, our number system uses the decimal point to separate whole numbers from numbers with a value *less than 1*.

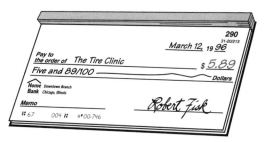

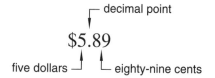

five dollars ⌐ ∟ eighty-nine cents

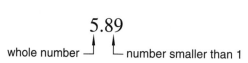

whole number ⌐ ∟ number smaller than 1

A digit's position in relation to the decimal point (its place value) tells you how large or small it is. Look at the **place value chart** below.

4 , 6 9 3 . 1 2 5

thousands hundreds tens ones tenths hundredths thousandths

> **Tip**
>
> Numbers *after* the decimal point end in *ths:* tenths, hundredths, thousandths, and so on.

What place is the 9 in? _____   What place is the 2 in? _____

You are correct if you said the 9 is in the *tens* place and the 2 is in the *hundredths* place.

Decimal values can be less than a thousandth. However, in this book and in many instances, you will not need to use these smaller values.

Understanding the Value of a Decimal

Example: Which is larger: 0.5 or 0.05?

Step 1
Look at the place value of the last digit in the decimal.

0.5
The place value is tenths.

0.05
The place value is hundredths.

Step 2
Think of a box divided into that many parts.

divided into tenths

divided into hundredths

Step 3
Visualize the whole decimal as part of that box.

5 tenths (0.5)

5 hundredths (0.05)

5 tenths is larger than 5 hundredths.

A. Use the place value chart on page 36 to fill in the blanks below.

1. Write the number *fifty-eight and one hundred sixty-seven thousandths.*

   Put a
   - 5 in the tens place
   - 8 in the ones place
   - 1 in the tenths place
   - 6 in the hundredths place
   - 7 in the thousandths place

   ____ ____ . ____ ____ ____

2. In the number 3,059.182, what digit is in the

   a. tens place? ____

   b. tenths place? ____

   c. ones place? ____

   d. hundredths place? ____

3. In the number 6,106.65, what digit is in the

   a. tens place? ____

   b. tenths place? ____

   c. ones place? ____

   d. hundredths place? ____

B. Choose the correct answer.

4. A library book has the number 791.48 on its spine. Which of the digits is in the tenths place?

   (1) 9

   (2) 4

   (3) 8

5. A security guard walks an average of 4.25 miles per day. What part of a mile does the digit 5 represent?

   (1) thousandths of a mile

   (2) hundredths of a mile

   (3) tenths of a mile

C. Four calculator displays are shown here. The fourth display is blank. Use the clues to figure out the contents of the fourth display.

Clue 1: There are no hundredths in the fourth display.

Clue 2: The fourth display has more tenths than the third display but fewer tenths than the first.

Clue 3: There are 4 times as many thousandths in the fourth display as there are in the third display.

6. **Explain** What is the number in the fourth display? How do you know?

Display 1 | 0.375 |

Display 2 | 0.046 |

Display 3 | 0.182 |

Display 4 | 0.____ |

Writing Decimals

When writing a decimal, be sure that

- a decimal point separates the whole number from the part smaller than 1
- the correct number of places follows the decimal point

Writing a Decimal

Example: Write "three and seventy-five thousandths" in numbers.

Step 1
Write the whole number. Replace the word *and* with a decimal point.

3.

↑
decimal point replaces *and*

Step 2
Decide how many places will follow the decimal point. Since this decimal value is *thousandths*, you need 3 places.

3.___ ___ ___

three places for thousandths

Step 3
If necessary, use zero as a placeholder. Write 75 in the last *two* of your three decimal places. Put a zero in the tenths place.

3.075

↑
zero as placeholder

Three and seventy-five thousandths is written **3.075**.

When there is no whole number, write a **leading zero** before the decimal point.

Example: *Four hundredths*

leading zero ⌐   ⌐ placeholder zero

0.04

A. **Write the following decimals in numbers. Replace *and* with a decimal point. Use zeros as placeholders if necessary.**

1. one and four tenths __.__

2. one and four hundredths __.__ __

3. twelve and eight tenths __ __.__

4. twelve and eight hundredths
   __ __.__ __

5. twelve and eight thousandths
   __ __.__ __ __

6. two and fifty hundredths __.__ __

7. two and fifty thousandths __.__ __ __

8. five hundred ten and seven hundredths
   __ __ __.__ __

9. five hundred ten and seven tenths
   __ __ __.__

10. five hundred ten and seven thousandths
    __ __ __.__ __ __

B. Choose the correct answer for each problem below.

11. A carpenter made a cut through a board.
The saw cut wasted one hundred twenty-five
thousandths of an inch. Which of the following
shows how much wood was wasted?

    (1) 1.25 inches

    (2) 0.125 inch

    (3) 0.0125 inch

12. A nurse walks an average of five and
thirty-five hundredths miles per day on
the job. Which of the following shows how
the distance is written in numbers?

    (1) 5.35 miles

    (2) 53.5 miles

    (3) 5.035 miles

13. The actual length of a year on Earth is
three hundred sixty-five and twenty-six
hundredths days. How would this amount
be written in numbers?

    (1) 365.026

    (2) 36,526

    (3) 365.26

14. A car company is offering car loans for as
low as two and nine tenths percent. What is
the loan rate written in numbers?

    (1) 2.9%

    (2) .29%

    (3) 2.09%

**C. You are recreating a landscaping sketch.
The measurements for the sketch are
given to you over the phone. Write the
measurements on the sketch shown here.
Be sure to replace *and* with a decimal
point. (The first one is started for you.)**

15. The distance from the birdbath to the rose
bush is *twelve and fifteen hundredths
centimeters.*

16. From the rose bush to the fence post is *eight
and one hundred twenty-five thousandths
centimeters.*

17. It is *twenty-two and five tenths centimeters*
from the fence post to the garage.

18. The distance from the garage to the birdbath
is *nineteen and six hundredths centimeters.*

19. It is *two and seven tenths centimeters* from the garage to the house.

20. **Discuss** Some people prefer to use the word *point* when reading decimals aloud.
For example, the distance 10.5 meters could be read as "ten point five meters"
instead of "ten and five tenths meters." Which method do you think is clearer?
Discuss your reasons with others.

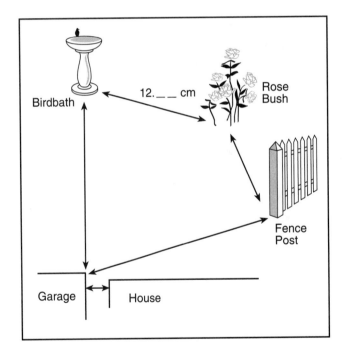

Decimals and Money

The U.S. money system is directly related to decimals. Dollars are written as whole numbers. Because there are 100 cents in a dollar, cents are written as hundredths.

Four dollars and seventy-eight cents

$ 4 . 78

Using Money Place Values

Example: Janice owes a customer $3.96 in change. She opens her register and finds she has only dollars, dimes, and pennies. How many of each will she give the customer?

You can use what you know about place value to solve this problem. You know the number 3.96 means 3 ones, 9 tenths, and 6 hundredths. Now replace the values with bills and coins.

$3 .	9	6
ones	tenths	hundredths
dollars	dimes	pennies

Janice could give the customer **3 dollars, 9 dimes, and 6 pennies.**

A. **Read the following amounts out loud. Then write them in words. Read the decimal point as "and" if it is preceded by a whole number. The first one is done for you.**

⌐ You don't need to read the "leading zero."

1. $0.85 *eighty-five cents* _____

2. $3.20 _____

3. $9.69 _____

4. $15.27 _____

5. $103.48 _____

6. $0.95 _____

7. $7.00 _____

8. $49.08 _____

B. Choose the correct answer for each problem.

9. Chun wrote a check for $105.09. How would you write this amount in words?

    (1) One hundred five nine dollars

    (2) One hundred and five dollars and ninety cents

    (3) One hundred five dollars and nine cents

10. Craig paid for lunch with a $5 bill. He received eighty-nine cents in change. How would the amount of change be written with numbers?

    (1) $0.89

    (2) $89

    (3) $089

11. Kina needs one dollar and thirty-five cents for bus fare. Which amount does she need?

    (1) $1.35

    (2) $0.135

    (3) $1.53

12. Nita has two hundred eight dollars and fifty-three cents in savings. How would the amount be written with numbers?

    (1) $280.53

    (2) $208.53

    (3) $285.30

C. Use the menu as needed to solve problems 13–17.

13. The pizza Mike ordered cost nine dollars and ninety-five cents. Which pizza did Mike order?

14. Florie ordered the vegetarian pizza. Write the cost of her pizza in words.

15. The cashier gave Florie two dollar bills, seven dimes, and five pennies. Write the amount of Florie's change in words and in numbers.

16. Omri has eight dollar bills and five dimes in his pocket. Can he afford a Hawaiian pizza? How do you know?

Anna's Pizzeria

Pizza "Original"	$6.70
Deluxe Pizza	$9.95
Hawaiian Pizza	$8.60
BBQ Chicken Pizza	$9.05
Vegetarian Pizza	$7.25
Individual Pizza	$4.00

17. **Write** You are figuring the cost of your order using a calculator. The display reads 10.7. Write what this is in dollars and cents.

Comparing Decimals

Because you use money every day, you know that $0.50 is more than $0.05. Fifty cents is more than 5 cents because 50 is greater than 5. When you take away the dollar sign, these decimals are written as 0.5 and 0.05, and 0.5 still has the greater value.

As you move farther to the right of the decimal point, the value of the digits gets smaller.

One tenth   **0.1**

The box below is divided into 10 equal parts. One tenth is shaded.

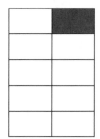

One hundredth   **0.01**

Each tenth in the previous box is divided into 10 equal parts. One hundredth is shaded.

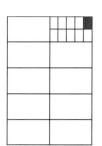

One thousandth   **0.001**

Imagine dividing each hundredth in the previous box into 10 equal parts and then shading only 1 of those parts. One thousandth is shaded.

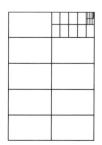

You can use what you know about money and place value to compare decimals.

Comparing Decimals

Example 1: Which is greater: 0.85 or 0.69?

Both numbers have the same number of decimal places. That is, both numbers are *hundredths*. Compare as you would with whole numbers.

0.85 is greater than 0.69 because 85 is greater than 69.

Example 2: Which is greater: 0.4 or 0.215?

Step 1
Add zeros so that both numbers have the same number of decimal places: *thousandths*.

0.400 ◀——— Adding zeros to the right of the last digit does not change the value of the decimal.

0.215

Step 2
Compare the decimals as if they were whole numbers.

400 is greater than 215, so 0.4 is greater than 0.215.

A. Compare the decimals below. Choose the larger decimal in each pair.

1. 0.6 or 0.82   0.15 or 0.09   0.325 or 0.411

2. 0.75 or 0.8   0.205 or 0.095   0.5 or 0.458

B. The symbols =, >, and < are used to make comparisons.
   Compare the numbers. Write the correct symbol in the blank.

Tip
The arrow always points to the smaller number.

   = is equal to   > is greater than   < is less than

3. 0.4 _____ 0.40   0.43 _____ 0.054   0.9 _____ 0.88

4. 0.002 _____ 0.015   0.9 _____ 0.95   0.25 _____ 0.075

C. An electronics firm produces electrical wire. The wires are color-coded.
   The diameters of eight types of wire are shown in the table. Use the
   information to solve problems 5–8.

5. Which color of wire has the smallest diameter?

6. Which color of wire has a diameter of *twenty hundredths* of an inch?

7. Which three colors have the greatest diameters?

8. **Explain** A technician needs a wire that is less than one tenth inch in diameter but greater than five hundredths inch. Which wires could be used? Explain how you know your answer is correct.

Micro Wire Products

Wire Insulation Color	Diameter in Inches
Red	0.06
Blue	0.15
Green	0.085
White	0.3
Yellow	0.025
Black	0.45
Brown	0.2
Violet	0.009

Adding Decimals

Adding decimals is like adding whole numbers. However, you need to line up the decimal points carefully.

Adding Decimals

Example: Mike plans to use a roasting bag to roast two pieces of beef. He needs to know the total weight of the meat to decide how long to cook it. According to the package, the larger piece weighs 4.8 pounds. The smaller weighs 3.3 pounds. What is the total weight of the meat?

Step 1	**Step 2**	**Step 3**	**Step 4**
Estimate first.	Line up the decimal points.	Bring the decimal point directly down in the answer.	Add. Regroup as needed.

4.8 ≈ 5	4.8	4.8	4.8
3.3 ≈ 3	+ 3.3	+ 3.3	+ 3.3
5 + 3 = 8		.	8.1

Check: Your answer of **8.1 pounds** is close to your estimate of 8 pounds.

To add some decimals, you may need to use **placeholder zeros** to line up the decimal points correctly. Adding zeros after the last digit following the decimal point does not change the number's value.

> **Tip**
> A whole number has an "understood" decimal point right after it.
> $$8 = 8. = 8.0 = 8.00$$

Using Placeholder Zeros

Example: Janna needs about 6 pounds of turkey for a recipe. She buys three packages of ground turkey at the grocery store. The packages weigh 2.75 pounds, 1.5 pounds, and 2 pounds. What is the total weight of the ground turkey?

Step 1	**Step 2**	**Step 3**	**Step 4**
Estimate first.	Line up the decimal points. Add placeholder zeros.	Bring the decimal point down in the answer.	Add. Regroup as needed.

2.75 ≈ 3	2.75	2.75	2.75
1.5 ≈ 2	1.50	1.50	1.50
	+ 2.00	+ 2.00	+ 2.00
3 + 2 + 2 = 7		.	6.25

Check: Your answer of **6.25 pounds** is close to your estimate of 7 pounds.

A. Add the following decimals. Write in placeholder zeros as needed. Line up the
decimals carefully in problem 2.

1.
```
    3.7            13.75          45.3             17
  + .05          + 36          +    .852        + 50.38
```

2. 1.4 + 3.07 + 27   6.8 + 3.5 + 9.6   0.13 + 39.6 + 5.28   0.05 + 4.95 + 9.5

B. Solve the problems below. You may use a calculator.

3. A customer bought items costing $2.74, $0.95, and $5.28. What was the total amount of the sale?

4. To make a recipe, Janet used two cans of beans. The small can contained 8.7 ounces, and the large can contained 15.5 ounces. How many ounces of beans did she use?

5. Three transistors weigh 0.516, 0.793, and 0.454 gram. What is the total weight of the transistors?

6. For lunch, Phil had a ham sandwich for $2.80, a cup of coffee for $0.75, and a piece of pie for $1.95. How much did Phil spend on lunch?

Making Connections: Using the Clear Entry Key

Entering numbers on a calculator can be tricky. There are many opportunities for keying in the wrong number.

Suppose you need to add four numbers. You enter the first three numbers and operation signs correctly, but you make a mistake when you enter the fourth number. What should you do? Do you have to start over?

No, you can simply clear the last entry. The Clear Entry key CE erases only what you see in the display. The rest of the numbers and operations are stored by the calculator. Press CE, then enter the number correctly and finish your calculation.

CE

The Clear Entry key clears only the last number entered.

Press this key once to clear the display. Press it twice to clear the entire calculation.

 Solve these addition problems using your calculator. If you enter a number wrong, press CE to clear the display.

1. 10.56 + 4.09 + 5.019   **2.** 3.15 + 2.005 + 6.95   **3.** $15.50 + $26.50 + $129.98

Subtracting Decimals

To subtract decimals, first line up the decimal points. Then subtract as you do with whole numbers.

Subtracting Decimals

Example: A mountain trail is 20.25 miles long. A hiker stops at the 12.8 mile marker to rest. How much farther does he have to hike?

Step 1
Estimate first.

$20.25 \approx 20$
$12.8 \approx 13$
$20 - 13 = 7$

Step 2
Line up the decimal points and add placeholder zeros.

$$\begin{array}{r} 20.25 \\ -\ 12.80 \\ \hline \end{array}$$

Step 3
Bring the decimal point directly down in the answer.

$$\begin{array}{r} 20.25 \\ -\ 12.80 \\ \hline . \end{array}$$

Step 4
Subtract. Regroup as needed.

$$\begin{array}{r} {}^{1\ 9}{}_{1} \\ 2\cancel{0}.25 \\ -\ 12.80 \\ \hline 7.45 \end{array}$$

Check: Your answer, **7.45 miles,** is close to your estimate of 7 miles.

A. Subtract the following decimals. Estimate an answer first.

Estimate

1.
$$\begin{array}{r} 5.4 \approx\ \ 5 \\ -\ 3.5 \approx -\ 3 \\ \hline \end{array}$$

$$\begin{array}{r} 3 \\ -\ .15 \\ \hline \end{array}$$

$$\begin{array}{r} 15.825 \\ -\ 9.95 \\ \hline \end{array}$$

$$\begin{array}{r} 3.056 \\ -\ 1.9 \\ \hline \end{array}$$

2. $9 - 6.75$ $24.5 - 16.9$ $1.078 - 0.36$ $3.4 - 2.705$

3. $8.2 - 7.9$ $12.85 - 5.87$ $7 - 5.352$ $42.6 - 21.92$

 B. To solve the problems below, subtract the second number from the first number. From that answer, subtract the third number. You may use a calculator.

4. $2.5 - 0.9 - 0.05$ $11.82 - 5.4 - 3.96$ $9 - 0.75 - 2.45$

5. $36 - 12.78 - 16.5$ $5.6 - 4 - 0.12$ $15.007 - 6.05 - 3.6$

C. Solve the problems below. Estimate an answer first.

6. A customer handed Connie a $20 bill to pay for $13.79 worth of merchandise. How much change did Connie give the customer?

7. Max's employer withholds $152.32 from Max's wages every two weeks. If Max earns $554 for two weeks of work, how much is his take-home pay for that period of time?

8. Robin ran the school's 100-meter dash in 12.4 seconds. The school record is 11.875 seconds. How much slower was Robin's run than the record?

9. Stuart had a 103.6° temperature at 6 P.M. After he took medicine, his temperature dropped to 99.8°. How many degrees did his temperature fall?

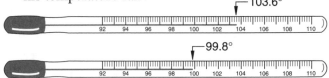

Making Connections: Keeping Mileage Records

The cost of some business travel is tax-deductible, so travelers often must keep mileage records. In a mileage record, you record the car's odometer reading at the end of each day. To find how many miles you have driven, you subtract yesterday's reading from today's odometer reading.

On the mileage record to the right, the number of miles driven for Monday was found by subtracting: 34,156.9 − 34,094.1 = 62.8.

Day	Odometer Reading	Miles Driven
Beginning Reading	34,094.1	
Monday	34,156.9	62.8
Tuesday	34,360.2	
Wednesday	34,615.1	
Thursday	34,704.9	
Friday	34,979.3	
Saturday	35,089.6	
Sunday	35,447.4	

Complete the Miles Driven column for the remaining days. Then solve the problems using the completed mileage record. You may use a calculator.

1. On which day did the traveler drive the farthest?

2. To find the total miles driven, you could add the numbers in the column labeled "Miles Driven." What other method could you use to find the total miles driven? What was the total number of miles driven for the week shown?

Solving Multistep Problems

Sometimes a problem involves more than one operation. On other problems you may need to use the same operation more than once.

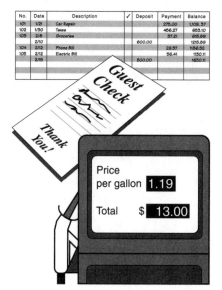

- You deposit money in your checking account. Then you write a check. How much money do you have in the bank? First *add* the deposit. Then *subtract* the check. The amount left is your balance.

- You order a meal at a restaurant. Then you use a $5-off coupon to reduce your bill. How much do you pay? First *add* the prices of the items you ordered. Then *subtract* the discount. The amount left is the amount you owe.

- On the last day of a trip, you have $36. You fill up the gas tank. Later, you stop for lunch. How much money do you have left? First *subtract* the cost of gas from $36. Then *subtract* the cost of lunch. The result is the amount you have left.

Solving Multistep Problems

Example: Jane is installing software on the hard drive of her computer. She needs 10 megabytes of storage, but she has only 4.6 megabytes free. She removes files to free up 3.8 more megabytes. How many more megabytes of storage does she need?

Step 1. Understand the question.	How many more megabytes does she need to get rid of to have 10 megabytes free?
Step 2. Decide what information is needed to solve the problem.	10 megabytes needed She has 4.6 free. She freed up 3.8 megabytes more.
Step 3. Think about how you might solve the problem.	To solve, I need two steps: 1. Add the amounts free now. 2. Subtract that number from the total she needs.
Step 4. Estimate an answer.	$5 + 4 = 9 \qquad 10 - 9 = 1$
Step 5. Solve the problem and check your answer. Does it make sense?	Use two operations: **Step 1** $\quad$ 4.6 $\qquad$ **Step 2** $\quad$ 10.0 $\qquad\quad \underline{+\ 3.8} \qquad\qquad\qquad \underline{-\ 8.4}$ $\qquad\qquad 8.4 \qquad\qquad\qquad\qquad 1.6$ *Check:* **1.6 megabytes** is close to your estimate of 1 megabyte and makes sense.

For each problem below, first decide if you need one or more steps to solve it. Write how you would solve it. Finally, solve the problem.

1. Dave's batting average at midseason was .277. By the end of the season, he had improved his average by .016. How short of batting .300 was he?

   a. Single-step or Multistep

   b. Write how you would solve it.

   _____

   _____

   c. Answer: _____

2. Using the information from the diagram below, what is the length of side A?

```
          ┌──────────┐
          │      5.4
   13.0    │         ┌────┐
          │         3.6 │
          │            │ A
          └────────────┘
```

   a. Single-step or Multistep

   b. Write how you would solve it.

   _____

   _____

   c. Answer: _____

3. Mrs. Smith's class is trying to raise money for a class trip to Sacramento. They estimate the trip will cost $6,690. They raise $373.90 at a bake sale. Next they raise $415.75 at a car wash. How much money have they raised so far?

   a. Single-step or Multistep

   b. Write how you would solve it.

   _____

   _____

   c. Answer: _____

4. Pat wants to travel from Los Angeles to Nashville. The trip will take 4 hours if she takes a nonstop flight. If she changes planes in Chicago, the trip will take longer. The flight from Los Angeles to Chicago will take 4.5 hours. Then she will have a 1.25 hour layover in Chicago. Finally, the flight from Chicago to Nashville will take 1.75 hours. How much time will she save by taking the nonstop flight?

   a. Single-step or Multistep

   b. Write how you would solve it.

   _____

   _____

   c. Answer: _____

Mixed Review

A. Choose each answer from the choices given.

1. The decimal 0.408 is read
   (1) forty-eight hundredths
   (2) four hundred eight hundredths
   (3) four hundred eight thousandths

2. In the decimal 0.75, the 7 is in the
   (1) ones place
   (2) tenths place
   (3) hundredths place

3. Which is a true statement?
   (1) 3.5 is equal to 3.50.
   (2) 3.5 is read "thirty-five hundredths."
   (3) 3.5 is greater than 3.55.

4. In the number 142.905, the zero is in the
   (1) tens place
   (2) tenths place
   (3) hundredths place

5. The amount of a check is $110.05. How is this amount written in words?
   (1) One hundred and ten dollars and ten cents
   (2) One hundred ten dollars fifty cents
   (3) One hundred ten dollars and five cents

6. Which is a true statement?
   (1) 0.078 is greater than 0.780.
   (2) 0.078 is less than 0.780.
   (3) 0.078 is equal to 0.780.

B. Solve the addition and subtraction problems below. You may use your calculator in problems 9 and 10.

7.
```
   12.4          $108.05          $35.62          9.006
 +  1.9         +  52.75         − 18.98         − 5.6
```

8. 56 + 18.75   4.06 − 2.5   30.5 + 5.88 + 14.9   4.5 − 1.125

9.
```
   1.006            5.8           $145.50         15.375
   0.904            3.2          − 89.98         − 6.5
 + 2.56          + 0.9
```

10. 2 − 0.55   3.5 + 0.98 + 6.75   12.9 − 5.09   15.06 + 8.67

C. Solve the problems below. Estimate each answer first.

11. Find the total length of the metal rods shown below.

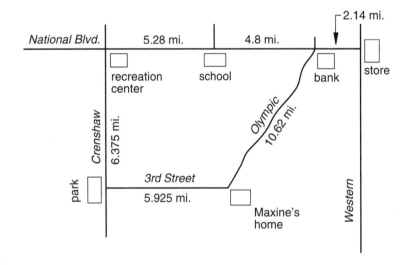

2.015 in.   3.875 in.   2.9 in.

Estimate: _____

Exact: _____

12. John is 1.925 meters tall. His 8-year-old son is 1.4 meters tall. How much taller is John than his son?

13. Tina purchased 11.4 gallons of gas on Monday. On Friday she bought 14.8 gallons. How many gallons did she buy altogether?

14. After the holiday sale, Byte Back raised the price of its Model NTX computer $299.95. During the sale, the computer was $1,825.75. What is the regular price of the computer?

15. For an assembly, Carrie needs a bolt measuring 0.935 inch in length. She has a bolt that measures 0.92 inch. Is the bolt too short or too long? By how much?

16. Ned has $204.50 in his checking account. How much will he have after making a $489.06 deposit and writing a check for $198.95?

D. Use the map to solve problems 17–20.

17. Using National Boulevard, how many miles is it from the recreation center to the store?

18. To drive to the school, Maxine can take Olympic to National, or she can take 3rd Street to Crenshaw to National.

   a. Which is the shorter route?

   b. By how many miles?

19. The school is almost halfway between the bank and the recreation center. By how many miles is the school closer to one than to the other?

20. Write Maxine needs to drop off her son at the park for soccer practice and then go to the bank. Describe the shortest route. How many miles is the shortest route?

Calculators and Decimals

$x - 7 = 3$

Calculators are useful for solving decimal problems because they express all numbers less than 1 in decimal form. As an experiment, use a calculator to divide 15 by 6.

Press 1 5 ÷ 6 =   Your display reads   2.5

When you divide 15 by 6, you get a remainder of 3.
The calculator always shows the remainder as a decimal.

$$2\ R3 = 15 \div 6 = 2.5$$
$$6\overline{)\ 15}$$
$$\underline{-\ 12}$$
$$3$$

Sometimes, the display will contain more decimal places than you need to solve the problem. Round your answer as needed.

Rounding Decimals

Example: Seven friends win a $67 prize at the office. If they divide the prize evenly, how much will each friend get?

Step 1
Estimate.

$67 rounds to $70
$70 ÷ 7 = $10

Step 2
Enter the problem on the calculator.

Press 6 7 ÷ 7 =

Step 3
Read the display.

9.5714285

Step 4
Since this decimal number represents money, round to the hundredths place.

9.5714285 → $9.57

Check: Your answer **$9.57** is close to the estimate of $10.

Some problems require more than one operation to find an answer. You can easily perform several operations on the calculator.

Multistep Problems on the Calculator

Example: Ellen bought 18 diskettes at $.83 each. She paid $1.08 in sales tax. What was the total amount of the sale?

Step 1
Estimate.

18 rounds to 20
$.83 rounds to $.80
$1.08 rounds to $1

20 × $.80 = $16
$16 + $1 = $17

Step 2
Enter the problem on the calculator.

Key in: The display reads:

1 8 ×   18.

. 8 3 +   14.94

1 . 0 8 =   16.02

Step 3
Read the display.

16.02

16.02 = $16.02

Check: Your answer **$16.02** is close to the estimate of $17.

A. Use your calculator to solve the problems below. In problems involving dollars and cents, round to the nearest cent.

1. $16.8 + 4.02$ 14.25×0.5 $21 \times \$0.12$

2. $1.053 - 0.79$ $\$15.69 + \$.89 + \$1.30$ $\$1,000 \div 15$

3. $\$3.75 \times 400 \times 3$ $\$20.50 \div 6$ $16.8 \times 7.25 \times 0.5$

B. Write the method for solving each problem. Estimate an answer. Then do the work on your calculator.

4. Max earns $8.96 per hour as a sales clerk. How much will he earn in 35.5 hours?

5. Barbara bought 15.8 gallons of gas at $1.19 per gallon. She also bought a quart of oil for $2.29. How much did she pay?

6. Peaches are $.36 each. Scott has $5. How many peaches can he buy? (*Hint:* You can't buy part of a peach.)

7. At a clothing store, men's shirts were reduced from $26.99 to $15.98. Mark bought 3 shirts. How much did he save off the original price?

Making Connections: The Value of a Remainder

Sometimes you need to find the value of a remainder.

Example: An airline sells 1,473 tickets for its new commuter flights. Each commuter plane can carry 55 passengers. If all but the last plane are full, how many passengers will be on the last plane?

Step 1
To find the number of flights, divide 1,473 by 55.

Your display reads

| 26.781818 |

Step 2
To find the total number of passengers on the full flights, multiply the whole number (26) by 55, the number you divided by.

$26 \times 55 =$

| 1430. |

Step 3
Subtract 1,430 from 1,473, the number you divided (the number of passengers in all).

$1,473 - 1,430 =$

| 43. |

There will be **43 passengers** on the last plane.

Use your calculator to solve the problems below.

1. Maggie has to box up 775 client files for storage. Each box holds 40 files. How many files will be left over to partially fill the last box?

2. The total cost of a car loan is $6,524. Colin agrees to make monthly payments of $182. How much will his final payment be?

Multiplying Decimals

Multiplying decimals is like multiplying whole numbers. You just need to place the decimal point correctly in your answer.

Multiplying a Decimal

Example: Tom is shipping 8 copies of a book. Each copy weighs 1.8 pounds. How many pounds will his shipment weigh?

Step 1	**Step 2**	**Step 3**	**Step 4**
Estimate first.	Line numbers up at the right.	Multiply as you would with whole numbers.	Count the digits after each decimal point in the problem. From the right, count out the same number of decimal places in the answer. Insert the decimal point.

1.8 rounds to 2	1.8	1.8	1.8	1 place
8 × 2 = 16	× 8	× 8	× 8	+ 0 place
		144	14.4 lb.	1 place

Check: Your answer, **14.4 pounds,** is close to the estimate of 16 pounds.

Sometimes when multiplying decimals, you must add a **placeholder zero** in your answer where it is needed to keep the correct number of decimal places.

Multiplying with Placeholder Zeros

Example: What is 0.05 × 0.9?

Step 1	**Step 2**	**Step 3**	**Step 4**
Estimate first.	Line numbers up at the right.	Multiply as you would with whole numbers.	Count the digits after each decimal point in the problem. From the right, count out the same number of decimal places in the answer. Insert the decimal point.

0.9 is nearly 1.	.05	.05	.05	2 places
1 × 0.05 = 0.05	× .9	× .9	× .9	+ 1 place
		45	.045	3 places

Add this placeholder zero to get 3 places after the decimal point.

Check: Your answer, **0.045,** is close to your estimate of 0.05.

Calculator Hint: Counting the number of decimal digits in the problem tells you the number of decimal digits that there should be in the answer. Use this fact to check your work when you multiply decimals on the calculator. If the display doesn't show the correct number of decimal digits, you probably entered a number wrong.

A. Put the decimal point in the correct place in each answer below.

1.
12.5	1 place
× 6.3	1 place
7875	2 places

3.4
× 5
170

.007
× .6
42

2.9
× .002
58

.19
× 6
114

B. Multiply the following decimals. Estimate first. Be careful where you put your decimal point in your final answer.

Estimate

2. 19 × .63   4.7 × 3.8   2.05 × 5   8.75 × 5.1

19	≈	20
× .63	≈	× .60

3. .8 × 3.5   7.7 × 1.9   3.25 × .25   60 × 9.9

C. Solve the problems below. You may use your calculator.

4. A roast weighing 4.5 pounds costs $4.40 per pound. What is the cost of the roast?

5. Pam's company pays her $0.31 for every mile she drives for work. On a business trip, she drove 575 miles. How much will her company pay her for her mileage?

6. Maggie bought 24 pencils at $.18 each. How much did the pencils cost?

7. Alan has a wooden dowel that is 100 centimeters in length. He needs to cut 11 pieces, each 8.25 centimeters in length, from the dowel. How many centimeters of dowel will be left over after he makes the cuts?

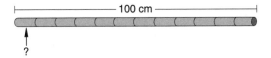

D. Read these shortcuts for multiplying a decimal by 10, 100, or 1,000. Then multiply the decimals below.

To multiply by 10, move the decimal point *1 place to the right.*

$2.6 × 10 = 2.6 = 26$

To multiply by 100, move the decimal point *2 places to the right.*

$2.6 × 100 = 2.60 = 260$
placeholder zero

To multiply by 1,000, move the decimal point *3 places to the right.*

$2.6 × 1,000 = 2.600 = 2,600$
placeholder zeros

8. 6.5 × 10   3.47 × 10   16.8 × 100   5.32 × 1,000

9. 10.05 × 100   3.032 × 10   1.125 × 1,000   .375 × 100

Dividing Decimals

When dividing a decimal by a whole number, be sure you place the decimal point in the correct place.

Dividing a Decimal by a Whole Number

Example: Andrew cut a 16.1-foot board into 4 equal pieces. How long is each piece?

Step 1
Estimate first.

16.1 rounds to 16.
16 ÷ 4 = 4

Step 2
Set up the problem. Place the decimal point directly above the decimal point in the dividend.

$$4\overline{)16.1}$$

divisor — | | — dividend

Step 3
Divide as you would whole numbers.

$$4\overline{)16.1} \quad \begin{array}{r} 4.0 \\ \underline{16} \\ 1 \end{array}$$

Step 4
Continue dividing. Add zeros if necessary and bring them down.

$$\begin{array}{r} 4.025 \\ 4\overline{)16.100} \\ \underline{16} \\ 10 \\ \underline{8} \\ 20 \\ \underline{20} \end{array}$$

Check: The answer **4.025** is close to the estimate of 4.

When dividing a number by a decimal, first change the divisor to a whole number. Then divide as shown above.

Dividing by a Decimal

Example: Marisol has a roll of lace that is 90 inches long. How many strips of lace 6.25 inches long can she cut from the roll?

Step 1
Estimate first.

90 rounds to 100.
6.25 is close to 5.
100 ÷ 5 = 20

Your estimate tells you to expect an answer that begins in the tens column.

Step 2
Move the decimal point in the divisor to the right until the divisor is a whole number.

$$6.25\overline{)90}$$

A whole number is understood to have a decimal point *after* it.

Step 3
Move the decimal point in the dividend *the same number of places.*

$$625.\overline{)9000.}$$

Add zeros so you can move the decimal point enough places.

Step 4
Bring the decimal point up from its new position. Divide as usual.

$$\begin{array}{r} 14.4 \\ 625.\overline{)9000.0} \\ \underline{625} \\ 2750 \\ 2500 \\ \underline{250\,0} \\ 250\,0 \end{array}$$

Check: Marisol could cut 14 strips of lace. Ignore the remaining 0.4 because the problem asks how many whole strips she can cut. You know you placed the decimal point correctly because your answer **14** is *reasonably* close to your estimate of 20.

A. Solve the following division problems.

1. $6\overline{)6.36}$ $3\overline{).075}$ $.4\overline{)3.4}$ $1.2\overline{)98.4}$

2. $1.792 \div 5.6$ $21.15 \div 9$ $.16 \div 32$ $6.84 \div .36$

B. Read these shortcuts for dividing a decimal by 10, 100, or 1,000.
   Then solve the division problems below.

To divide by 10, move the decimal point *1 place to the left.*

$13.5 \div 10 = 1\underset{\smile}{3}.5 = 1.35$

To divide by 100, move the decimal point *2 places to the left.*

$13.5 \div 100 = \underset{\smile}{13}.5 = 0.135$

To divide by 1,000, move the decimal point *3 places to the left.*

$13.5 \div 1,000 = \underset{\smile}{013}.5 = 0.0135$

Add zeros as necessary.

3. $15.72 \div 10$ $2,058 \div 1,000$ $25.5 \div 10$ $0.06 \div 100$

C. Solve the following word problems. You may use a calculator.

Problems 4 and 5 refer to average.

To find an **average,** *add* a group of numbers, then *divide* the total by the number of items in the group.

Example: Find the average of 2.6, 4.6, and 1.5.

$2.6 + 4.6 + 1.5 = 8.7$

$8.7 \div 3 = 2.9$ The average is 2.9.

4. Rainfall for the last 3 months was 6.2, 8.5, and 3.9 inches. What was the average rainfall per month for the 3-month period?

5. Craig ran 4 days last week. His distances were 8.25, 6.5, 7.25, and 10.5 kilometers. What was his average distance per day for the 4 days?

6. Four concert tickets cost $154.68. How much does each ticket cost?

7. One serving of potato chips is 1.75 ounces. The entire can of chips holds 15.75 ounces. How many servings are in the can?

1 serving = 1.75 oz.

Use the table for problems 8 and 9.

Monday	7.75 hours
Tuesday	6.5 hours
Wednesday	7.25 hours
Thursday	8.0 hours
Friday	7.25 hours

8. Evan's work schedule for next week is shown above. What is the average number of hours he is scheduled to work per day?

9. **Explain** Evan's gross pay (before deductions) for the week will be $338.10. How much does he earn per hour? Explain.

Choosing a Method

Some test questions ask you to choose the best way to solve a problem. Instead of calculating an answer, you choose a **numerical expression** that can be used to compute the answer.

Choosing Numerical Expressions

Example: Marci bought 6 donuts for $.55 each. She paid $.24 tax. Which expression can be used to find how much Marci spent?

(1) 6(0.55)(0.24)

(2) 6(0.55) + 0.24

(3) 6 + 0.55 + 0.24

Tip

When parentheses are used to separate numbers, you don't need to write a multiplication sign. 6(0.55) means 6 × 0.55

To solve the problem, you would *multiply* the number of donuts by the price, then *add* the sales tax. Only choice **(2)** does that.

Parentheses can also be used to show that a certain operation must be done first. Notice how the values of these expressions that use the same numbers and operations change because of the parentheses.

3(4) + 5 =   3(4 + 5) =

12 + 5 = 17   3(9) = 27

These guidelines, called the **order of operations,** will help you read and understand numerical expressions:

- First, look for any operations in parentheses. These should always be done first.
- Second, working from left to right, find any multiplication and division steps. These are done next.
- Third, working from left to right, find any addition and subtraction steps. These are done last.

A. **Match the numerical expressions to the descriptions.**

a. 12(5) + 15 − 7   **b.** 7(12 + 5)   **c.** (15 − 7) + (12 − 5)   **d.** $\frac{(12 - 5)}{7}$ *Hint:* A fraction bar is used to show division.

1. Subtract 7 from 15. Then subtract 5 from 12. Add the differences. _____

2. Multiply the sum of 12 and 5 by 7. _____

3. Multiply 12 by 5. Then add 15. Subtract 7 from the total. _____

4. Subtract 5 from 12. Then divide by 7. _____

B. Choose the best method to solve each problem. Do not solve.

5. A builder bought 80 lighting fixtures at $39.50 each. She paid $252.80 in sales tax. Which expression shows how to find the total amount the builder spent?

   (1) 80($252.80) + $39.50

   (2) $80 + $252.80 + $39.50

   (3) 80($39.50) + $252.80

6. The pollution index readings in Los Angeles for 3 days were 45.9, 42.9, and 53.7. Which expression could be used to find the average daily reading for the 3 days?

   (1) $45.9 + 42.9 + \frac{53.7}{3}$

   (2) $3(45.9 + 42.9 + 53.7)$

   (3) $\frac{(45.9 + 42.9 + 53.7)}{3}$

7. The Suttons chose carpeting that costs $12.69 per square yard. They need 20 square yards for one room and 15.8 square yards for another. Which expression can be used to find the cost of carpeting both rooms?

   (1) $12.69(20 + 15.8)

   (2) $12.69 + 20 + 15.8

   (3) $12.69(20) − $12.69(15.8)

Problem 8 refers to the diagram.

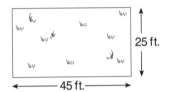

8. An outdoor dog show needs to put up temporary fencing to form a rectangle measuring 45 feet by 25 feet. Which expression can be used to find the number of feet of fencing needed?

   (1) 2(45 + 25)

   (2) 2(45) + 25

   (3) 2(45)(25)

For another look at formulas, turn to page 271.

Making Connections: Finding Another Way

Sometimes the method you would use to solve the problem isn't among the choices. What can you do? Find another way to write the solution.

Look at these two expressions: 3(6 + 8)   3(6) + 3(8)

Both show different ways of writing the same thing. They are both equal to 42. Solve them to see if this is true.

Write another expression equal to each of the following expressions.

   1. 10(25 − 15)   **2.** 4(8) + 4(2)   **3.** 2(16 + 9)

Figuring Unit Price and Total Cost

We use unit pricing to find better buys. Would you rather buy 3 apples for $1 or pay $.50 per apple? At 3 for $1, you are paying only about $.33 per apple. This is the best buy. **Unit pricing** means finding the cost of one item or unit of measurement.

Finding Unit Price

Example: A 16-fluid-ounce can of juice costs $1.40. What is the price per fluid ounce?

Step 1	**Step 2**	**Step 3**
Estimate.	Divide the price by the number of units.	Round money amounts to the nearest hundredth.
$1.40 is close to $1.60.		
$1.60 ÷ 16 = $.10	$1.40 ÷ 16 = $.0875	$.0875 rounds to $.09

Check: Your answer **$.09** is close to your estimate of $.10.

 A. Find the unit price for each product. You may use a calculator.

1. A store display reads: What is the price of 1 bar?

3. Speedy Copy offers business customers 100 copies for $5.50. What is the unit price of one copy?

2. At a produce stand, tomatoes are 5 for $2. What is the price of 1 tomato?

4. If 3 hamburgers cost $3.87, what is the unit price of 1 hamburger?

Sometimes you have the unit price, but you need to find the **total cost** for the number of units you want to buy.

Finding Total Cost

Example: A caterer charges $6.75 per guest. The host of a party has invited 32 guests. How much does he owe the caterer?

Step 1
Estimate.

$6.75 rounds to $7
32 rounds to 30

30 × $7 = $210

Step 2
Multiply the unit price, the amount per guest, by the number of guests.

```
   $6.75
 ×    32
   13 50
  202 5
 $216.00
```

Check: Your answer of **$216.00** is close to your estimate of $210.00.

B. Use the unit price to find the total cost for each product. You may use a calculator.

5. Extra-lean ground beef is $3.49 per pound. How much will 3 pounds cost?

6. Dinner rolls are 14 cents each at a bakery. How much will a customer be charged for 2 dozen rolls?

7. Super unleaded gasoline is $1.49 at the full-serve pump. How much will 18.5 gallons cost to the nearest cent?

8. At a grocery co-op, rice is sold by the pound. One pound is $.69. How much will a 5-pound bag of rice cost?

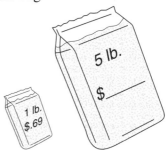

C. The ads for two grocery stores are shown below. Both have the same products. For each item, find the unit price at each store. Then write the name of the store that offers the best buy. You may use a calculator.

—— A & R Markets ——

Lucky Cereal .... 32 oz. box $1.92

Boneless Ham .... 3 lb. for $12.75

Toilet Paper .... 6 rolls for $2.04

Apples .... 2 lb. for $1.00

Baby Shampoo .... 12 fl. oz. for $2.88

Foodtime

Lucky Cereal .... 18 oz. box $1.26

Boneless Ham .... 4 lb. for $16.40

Toilet Paper .... 9 rolls for $2.52

Apples .... 5 lb. for $2.75

Baby Shampoo .... 8 fl. oz. for $2.08

	A & R Unit Price	Foodtime Unit Price	Best Buy
Lucky Cereal			
Boneless Ham			
Toilet Paper			
Apples			
Baby Shampoo			

9. Write What factors besides price might convince customers to shop at a particular store? Can you think of any way that math could be used to measure these factors?

Unit 2 Review

A. Solve the problems below. Estimate an answer first. You may use a calculator to do problems 4 and 5.

1. 4.95 − 2.6   190.2 − 9.95   10.04 − 8

2. 5.375 + 3.125   17.9 + 18.21   1.035 + 6.98

3. 4.9 × 2.07   3.25 × 8   50.5 × 4.08

4. 10 ÷ 2.5   21.84 ÷ 52   542 ÷ 5.42

5. 3.2(5 + 6.5)   102 ÷ (100 − 85)   (2.8 × 6) + (1.5 × 8)

B. Choose the best answer to each word problem.

6. The Mataele family is financing a new refrigerator. The family agrees to make a $250 down payment and 24 monthly payments of $39.52. Which expression can be used to find the total amount the Mataeles will pay for the refrigerator?

(1) $250 + 24 + $39.52

(2) $250 + 24($39.52)

(3) 24($250) + 24($39.52)

(4) 24($250 + $39.52)

(5) 24($250 − $39.52)

7. Bill needs to buy diskettes for his computer. Brand A has 12 in a box for $7.68. Brand B has 10 in a box for $6.40. Which is a true statement about the disks?

(1) Bill cannot find the unit price of either brand.

(2) The unit price for both brands is the same.

(3) Brand A is the better buy.

(4) Brand B is a slightly better buy.

(5) Brand B is a much better buy.

8. Put the following weights in order from *greatest* to *least*:

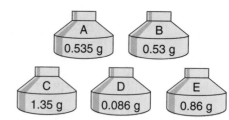

(1) C, E, A, B, D

(2) A, B, D, E, C

(3) A, C, D, E, B

(4) C, D, A, B, E

(5) C, A, D, E, B

9. Four roommates split the cost of a utility bill evenly. If the bill is $87.69, how much is each person's share to the nearest penny?

(1) $20.25

(2) $20.93

(3) $21.22

(4) $21.92

(5) $21.25

10. Fran has $35.00 to spend on art supplies. Tubes of paint are $1.72 each. How many tubes of paint can she buy?

(1) 2

(2) 17

(3) 19

(4) 20

(5) 23

11. The cost to a restaurant for a pie is $5.25. The pie is cut into 8 slices, which are sold for $1.95 each. Which expression can be used to find the profit the restaurant makes if all slices are sold?

(1) 8($1.95) − $5.25

(2) $5.25 − 8($1.95)

(3) $5.25 + $1.95

(4) 8($5.25 − $1.95)

(5) 8($5.25 + $1.95)

C. This portion of a check register belongs to Kate Mackin. Use the check register to solve problems 12–15. You may use a calculator.

No.	Date	Transaction	Payment (−)	✓ (T)	Deposit (+)	Balance
						97.41
135	10/9	Department of Water and Power (utilities)	37.18			60.23
	10/15	Deposit (paycheck)			263.90	324.13
136	10/15	Food King Market (groceries)	56.91			267.22
137	10/16	U-Save Auto (gasoline)	15.50			
	10/22	Deposit (check from mom for birthday)			50.00	

12. What is Kate's balance after the calculations involving check #137 and the $50.00 deposit are completed?

13. Check #136 shows Kate's grocery bill for 1 week. In October she bought groceries 3 other times, spending $68.90, $72.20, and $61.75. What was her average weekly grocery bill for October?

14. Kate needs to write a $400 check for her rent. What is the smallest amount she would need to deposit into her account to cover the check and leave a $20 balance?

15. Kate gets paid 52 times per year. The deposit on 10/15 shows the take-home pay for 1 pay period. How much does Kate take home in a year?

Working Together

A baseball player's batting average shows how many hits the player has compared to the number of times at bat. The average is found by dividing the number of hits by the number of at-bats. The answer is carried out to the thousandths place. Working in a small group, think of other sports statistics you could calculate.

Unit 3

Fractions, Ratios, and Percents

Skills

Relating decimals, fractions, ratios, and percents

Different forms of fractions

Equivalent fractions

Adding and subtracting fractions

Finding common denominators

Multiplying and dividing fractions

Writing ratios and proportions

Solving problems with proportions

Understanding percents

Writing and solving percent equations

Tools

English and metric rulers

Problem Solver

Two-step percent problems

Applications

Distances

Discounts

Like decimals, **fractions, ratios,** and **percents** can be used to show part of a whole. We use these fractional systems every day.

A recipe calls for $\frac{1}{2}$ cup of sugar (fraction). The newspaper reports that 48% of all registered voters voted in the last election (percent). A television commercial boasts that children prefer Tasty Flake cereal 9 to 1 over the leading brand (ratio).

In each example, a whole object or group is divided into parts. The fraction represents a part of the whole.

In this unit, you will learn the meanings and uses of these fractional systems. You will also see how decimals are related to fractions, ratios, and percents. This will help you solve many problems.

When Do I Use Fractions, Ratios, and Percents?

Each of the following situations uses fractions, ratios, or percents. Check the experiences you have had.

☐ following a recipe

☐ paying interest on a loan

☐ measuring with a ruler, yardstick, or tape measure

☐ figuring out the amount of a discount at a store sale

☐ reading a circle graph in the newspaper

☐ comparing win-loss records of teams

☐ figuring out how much to leave for a tip

☐ figuring what part of your income goes to taxes

If you've done any of those things, you've used these fractional systems. Describe below some of the experiences you've had.

1. Have you ever bought anything on credit? How did you know how much interest you were paying?

_____

_____

2. Have you ever used a ruler or yardstick to measure length? What fractional parts was the ruler or yardstick divided into?

_____

_____

3. When you eat at a restaurant, how do you figure out how much to leave for the tip? Suppose you want to leave a 15% tip. How could you figure out the amount?

_____

_____

Talk About It

Have you ever had to estimate the size of a room? How can you estimate feet without a ruler or yardstick? Suppose you had to guess the distance a golfer hits a golf ball. How can you estimate yards? Discuss your methods with others.

🖩 **Throughout this unit,** you may use a calculator wherever that would be useful.

Relating Decimals and Fractions

A **fraction** represents a part of a whole. Decimals are a kind of fraction because they show parts of a whole. For example, fifty cents ($.50) is a fraction of one dollar ($\frac{1}{2}$ dollar).

Decimals use place value to show the value of a digit. The decimal 0.7 means *seven tenths* because the digit 7 is in the tenths place. You can also write *seven tenths* as $\frac{7}{10}$.

The shaded portion of this figure can be written 0.7 or $\frac{7}{10}$. Both are read "seven tenths."

In the fraction $\frac{7}{10}$, the top number, or **numerator,** tells how many parts are shaded. The bottom number, the **denominator,** tells how many total parts there are in the figure.

$$\frac{7}{10} \quad \frac{\text{numerator}}{\text{denominator}}$$

A. Write a decimal and a fraction to represent the shaded portion of each figure.

1.

Decimal _____

Fraction _____

2.

Decimal _____

Fraction _____

3.

Decimal _____

Fraction _____

When the denominator of a fraction is 10, 100, 1,000, or another multiple of 10, you can easily change the fraction to a decimal.

Changing Fractions to Decimals by Using Place Value

Example: Write $\frac{75}{1,000}$ as a decimal.

Step 1
Think: How many decimal places does a fraction with a denominator of 1,000 need? It needs 3 decimal places.

0.__ __ __
      ⌐ thousandths place

Step 2
The numerator should fill the place value named by the denominator. Use placeholder zeros if necessary.

0.075 means "seventy-five thousandths"
  ⌐ placeholder zero

B. Write these fractions as decimals.

4. $\frac{69}{100}$ $\frac{5}{10}$ $\frac{482}{1,000}$ $\frac{3}{100}$

5. $\frac{2}{10}$ $\frac{25}{100}$ $\frac{1}{10}$ $\frac{9}{1,000}$

C. Use your knowledge of decimals and fractions to compare each pair of numbers. Write = (is equal to), > (is greater than), or < (is less than) in each blank.

6. $\frac{50}{100}$ _____ 0.48   0.6 _____ $\frac{6}{10}$ 0.035 _____ $\frac{350}{1,000}$

7. $\frac{5}{1,000}$ _____ 0.005 $\frac{16}{100}$ _____ 0.20   0.9 _____ $\frac{8}{10}$

D. Choose the best answer for each problem below.

8. In a race, a speed skater shaved $\frac{8}{100}$ of a second off the course's best lap time. Which decimal represents the time saved?

    (1) 0.8

    (2) 0.08

    (3) 0.008

9. The Utvich family spends 0.3 of their income on rent. What fraction of their income is spent on rent?

    (1) $\frac{3}{10}$

    (2) $\frac{3}{100}$

    (3) $\frac{3}{1,000}$

10. In a recent survey, 85 out of every 100 people asked said that they eat fast food at least twice a week. Which decimal represents this group?

    (1) 0.0085

    (2) 0.085

    (3) 0.85

ProCorp Assembly	
Shift	**Defect Rate**
Morning	6/1,000
Evening	1/100
Swing	6/100

11. Which assembly line shift had a defect rate of 0.06?

    (1) the morning shift

    (2) the evening shift

    (3) the swing shift

12. Explain Did the morning or the swing shift make errors at a lower rate? Explain how you know which fraction is smaller.

13. Which shift had the lowest defect rate?

Different Forms of Fractions

Fractions represent a part of a whole. They can also represent division.

The fraction $\frac{1}{4}$ can mean "1 out of 4 parts."

$\frac{1}{4}$ can also mean "1 divided by 4." The fraction bar in a fraction means "divided by."

$\frac{1}{4}$ | **1** | **2** | **3** | **4**

$\frac{1}{4}$ of 4 is 1.

$\frac{1}{4}$

1 divided by 4 is $\frac{1}{4}$ of a whole.

A fraction with the top number equal to or greater than the bottom number is called an **improper fraction.**

▶ When the numerator of a fraction is the *same* as the denominator, the fraction is *equal to 1.* $\frac{4}{4} = 1$ because 4 divided by 4 is 1.

Write a fraction to express the shaded part of this figure:

 $\dfrac{\text{number of shaded parts}}{\text{number of total parts}}$

The correct fraction, $\frac{3}{3}$ is the same as **1.** In other words, the whole figure is shaded. Expressed as a division problem, $3 \div 3 = 1$.

▶ When the numerator of a fraction is *greater* than the denominator, the fraction is *greater than 1.* $\frac{8}{5} > 1; \frac{9}{3} > 1$

Write a fraction to express the shaded portions of these figures:

 $\dfrac{\text{number of shaded parts}}{\text{number of parts each square is divided into}}$

The correct fraction is $\frac{3}{2}$. Each square is divided into 2 parts. Three parts are shaded. $\frac{3}{2}$ is greater than 1. We can treat this fraction as a division problem:

$$\frac{3}{2} = 2\overline{)3}^{\,1\,R1} = 1\frac{1}{2}$$ Put the remainder 1 over the original denominator 2. $\frac{3}{2} = 1\frac{1}{2}$

The number $1\frac{1}{2}$ is called a **mixed number** because it is a "mix" of a whole number and a fraction. Dividing a fraction greater than 1 converts it to a mixed number or a whole number.

A. Write each fraction as a whole number or as a mixed number. The first one has been done for you.

1. $\frac{7}{5} = 7 \div 5 = 1\frac{2}{5}$ $\qquad\qquad \frac{4}{3} \qquad\qquad\qquad \frac{11}{2} \qquad\qquad\qquad \frac{6}{6}$

2. $\frac{6}{3}$ $\qquad\qquad\qquad\qquad \frac{12}{7} \qquad\qquad\qquad \frac{5}{4} \qquad\qquad\qquad \frac{7}{3}$

Changing Mixed Numbers to Fractions

To write a fraction as a mixed number, you divide. To write a mixed number as a fraction, you do the opposite—multiply.

Writing a Mixed Number as a Fraction

Example: Write $2\frac{1}{4}$ as a fraction.

Step 1
Multiply the whole number by the denominator. Write this product over the denominator of the fraction.

$2\frac{1}{4} \quad \frac{8}{4} \longleftarrow \quad 2 \times 4 = 8$

Step 2
Add the original numerator to the product found in Step 1. Write the total over the denominator.

$2\frac{1}{4} = \frac{8}{4} + \frac{1}{4} = \frac{9}{4}$

B. Change these mixed numbers into fractions.

3. $4\frac{1}{2}$ $\qquad\qquad 3\frac{1}{5} \qquad\qquad\qquad 1\frac{1}{3} \qquad\qquad\qquad 4\frac{3}{8}$

4. $2\frac{2}{3}$ $\qquad\qquad 1\frac{3}{5} \qquad\qquad\qquad 3\frac{5}{6} \qquad\qquad\qquad 5\frac{3}{4}$

Making Connections: Calculators and Fractions

Calculators use decimals to represent fractional amounts. To work fraction problems on a calculator, change the fractions to decimals.

Example: How can Marco enter the fraction $\frac{3}{4}$ on his calculator?

Remember, the fraction bar means "divided by." $\frac{3}{4}$ means $3 \div 4$. Use the calculator to divide 3 by 4. The display will read 0.75. $\frac{3}{4}$ = **0.75**

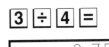

Use a calculator to change these fractions to decimals.

1. $\frac{3}{5}$ $\qquad$ **2.** $\frac{7}{8}$ $\qquad$ **3.** $\frac{1}{2}$ $\qquad$ **4.** $\frac{1}{8}$ $\qquad$ **5.** $\frac{3}{4}$ $\qquad$ **6.** $\frac{2}{5}$

Equivalent Fractions

Write a fraction to express the shaded portion of each figure.

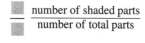 $\dfrac{\text{number of shaded parts}}{\text{number of total parts}}$

 $\dfrac{\text{number of shaded parts}}{\text{number of total parts}}$

Compare the two shaded areas. What do you notice?

The shaded parts of the two figures above are the same size. $\frac{3}{4}$ and $\frac{6}{8}$ are **equivalent fractions** because they are equal to each other. The fractions represent equal amounts.

$$\frac{3}{4} = \frac{6}{8}$$

▶ To write an equivalent fraction, you can *multiply* the numerator and the denominator by the same number. This is the same as multiplying by 1.

Example: $\frac{1}{4} \times \frac{2}{2} = \frac{2}{8}$

means 2 ÷ 2 or 1

Multiplying by 1 does not change the value of a number. Since $\frac{2}{2} = 1$, you are not changing the *value* of $\frac{1}{4}$ when you multiply by $\frac{2}{2}$. You are changing the *form* of the fraction. $\frac{1}{4} = \frac{2}{8}$

▶ To write an equivalent fraction, you can also *divide* both the numerator and denominator by the same number. This is the same as dividing by 1.

Example: $\frac{8}{12} \div \frac{4}{4} = \frac{2}{3}$

means 4 ÷ 4 or 1

Dividing by 1 does not change the value of a number. Since $\frac{4}{4} = 1$, you are not changing the *value* of $\frac{8}{12}$ when you divide by $\frac{4}{4}$. You are changing the *form* of the fraction. $\frac{8}{12} = \frac{2}{3}$

▶ A fraction is in **lowest terms** when both the numerator and denominator are as small as they can be. In other words, they cannot be evenly divided by the same amount.

Example: Is $\frac{3}{6}$ written in lowest terms?

No. Both numerator and denominator can be evenly divided by 3. $\frac{3}{6}$ written in lowest terms is $\frac{1}{2}$.

$$\frac{3}{6} \div \frac{3}{3} = \frac{1}{2}$$

A. Write equivalent fractions for the fractions below. Multiply the fractions in problem 1 by $\frac{2}{2}$. Multiply the fractions in problem 2 by $\frac{3}{3}$.

1. $\frac{3}{5} \times \frac{2}{2} =$ _____ $\frac{1}{6} \times$ ___ $=$ ___ $\frac{5}{8} \times$ ___ $=$ ___ $\frac{1}{3} \times$ ___ $=$ ___

2. $\frac{1}{4} \times$ ___ $=$ ___ $\frac{3}{8} \times$ ___ $=$ ___ $\frac{9}{10} \times$ ___ $=$ ___ $\frac{3}{7} \times$ ___ $=$ ___

B. Write equivalent fractions by dividing each fraction in problem 3 by $\frac{2}{2}$ and each fraction in problem 4 by $\frac{3}{3}$.

3. $\frac{8}{14} \div \frac{2}{2} =$ _____ $\frac{4}{10} \div$ ___ $=$ ___ $\frac{4}{16} \div$ ___ $=$ ___ $\frac{12}{24} \div$ ___ $=$ ___

4. $\frac{6}{9} \div$ ___ $=$ ___ $\frac{15}{18} \div$ ___ $=$ ___ $\frac{21}{27} \div$ ___ $=$ ___ $\frac{9}{30} \div$ ___ $=$ ___

C. Are these fractions in lowest terms? If yes, write the fraction in the space provided. If not, put the fraction in lowest terms.

5. $\frac{4}{5} =$ _____ $\frac{12}{18} =$ _____ $\frac{2}{8} =$ _____ $\frac{5}{7} =$ _____

6. $\frac{20}{24} =$ _____ $\frac{16}{30} =$ _____ $\frac{4}{6} =$ _____ $\frac{3}{12} =$ _____

D. MedLab Manufacturing makes rubber tubing with the diameters shown on the chart. Use the chart to solve the problems.

MedLab Tubing

Model Number	Diameter
#4AB	$\frac{1}{32}$ in.
#4C	$\frac{1}{16}$ in.
#5D	$\frac{1}{4}$ in.
#5E	$\frac{5}{16}$ in.
#6F	$\frac{3}{8}$ in.
#6GH	$\frac{1}{2}$ in.

7. Ae Ri calculates that she needs tubing with a diameter of $\frac{8}{32}$ inch. Which model of tubing should she order?

8. A hospital has a large stock of Model #5E on hand. If tubing with a diameter of $\frac{20}{64}$ inch is needed, is Model #5E too large, too small, or just right?

9. Explain A company supervisor suggests that the tubing list be rewritten so that every measurement is listed in thirty-seconds of an inch. How should the measurement for Model #6GH be written? Explain how you know your answer is correct.

English and Metric Rulers

A common measuring tool found in many homes is the **English ruler** (shown here).
It is marked in inches (in.) and fractions of an inch.

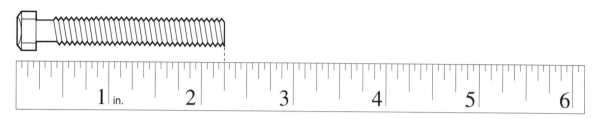

Each inch is divided into $\frac{1}{16}$-, $\frac{1}{8}$-, $\frac{1}{4}$-, and $\frac{1}{2}$-inch fractions. Different fraction units
are shown by using marks of different lengths. For example, $\frac{1}{2}$-inch marks are
longer than $\frac{1}{4}$-inch marks.

The bolt's length is $2\frac{1}{4}$ inches. If you count the eighth- or sixteenth-inch marks, its
length would be $2\frac{2}{8}$ or $2\frac{4}{16}$ inches. In lowest terms, both lengths equal $2\frac{1}{4}$ inches.

$$2\frac{2}{8} = 2\frac{1}{4}$$
$$2\frac{4}{16} = 2\frac{1}{4}$$

A **metric ruler** (shown below) is marked in centimeters (cm) and millimeters (mm).
Each centimeter is divided into 10 millimeters.

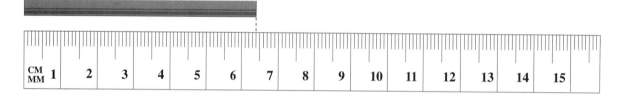

You can write the length of the metal rod three ways:

- *Millimeters only:* 64 mm
- *Centimeters and millimeters:* 6 cm 4 mm
- *Centimeters only:* 6.4 cm

$$64 \text{ mm} = 6 \text{ cm } 4 \text{ mm} = 6.4 \text{ cm}$$

Since the metric system is based on units of 10 like the decimal system,
millimeters can be written in decimal form as a fraction of centimeters.

$$4 \text{ mm} = .4 \text{ cm}$$

A. Write the lengths of the objects below.

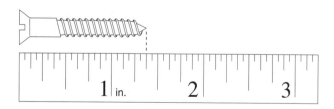

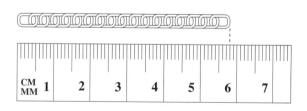

1. _____ in.

2. _____ cm _____ mm or _____ cm

3.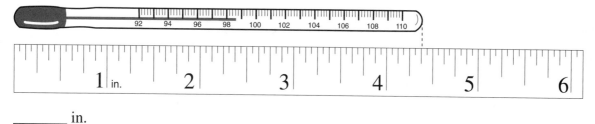

_____ in.

4.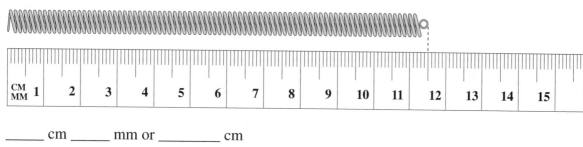

_____ cm _____ mm or _____ cm

B. How far from the left end of the ruler is each point shown below?

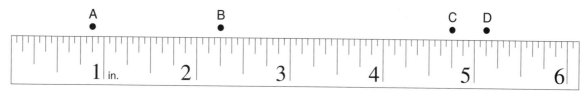

5. Point A = _____ Point B = _____ Point C = _____ Point D = _____

C. Home Builder sells screws and bolts according to length. The current display is shown below. Use the display and your knowledge of measurement to answer the questions.

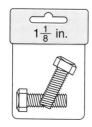
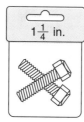
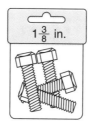
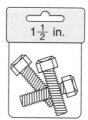
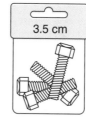
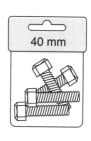

$1\frac{1}{8}$ in. $1\frac{1}{4}$ in. $1\frac{3}{8}$ in. $1\frac{1}{2}$ in. 3.5 cm 40 mm

6. For a repair call, Jim needs a bolt that is at least $1\frac{3}{16}$ inches long. He wants to use the shortest possible bolt that will do the job. Of those shown, which length of bolt should he buy?

7. Explain Aubry needs a bolt that is 35 millimeters long. Of those shown, which package should she buy? Explain how you arrived at your answer.

Adding and Subtracting Like Fractions

Like fractions are fractions that have the same denominator. For example, $\frac{1}{8}$ and $\frac{3}{8}$ are like fractions. $\frac{1}{8}$ and $\frac{3}{4}$ are **unlike fractions** because they have different denominators. To add like fractions, just add the numerators. The denominator remains the same.

$$\frac{1}{8} + \frac{3}{8} = \frac{4}{8} \longleftarrow 1 + 3 = 4$$

The answer to an addition or subtraction problem may need to be simplified. Always write your answers in lowest terms. Change improper fractions to mixed numbers.

Adding Like Fractions

Example: $\frac{5}{8} + \frac{4}{8}$

Step 1
Add the numerators.

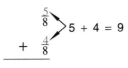

$5 + 4 = 9$

Step 2
Place the result over the denominator.

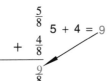

$5 + 4 = 9$

Step 3
Change to a mixed number if necessary.

$$\frac{9}{8} = 8\overline{)9}^{\,1\,R1} = 1\frac{1}{8}$$

Turn to page 68 for a review of changing improper fractions to mixed numbers.

Subtracting Like Fractions

Example: $\frac{3}{4} - \frac{1}{4}$

Step 1
Subtract the numerators.

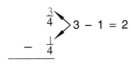

$3 - 1 = 2$

Step 2
Place the result over the denominator.

$$\frac{3}{4}$$
$$-\frac{1}{4}$$
$$\frac{2}{4}$$

$3 - 1 = 2$

Step 3
Simplify if necessary.

$$\frac{3}{4}$$
$$-\frac{1}{4}$$
$$\frac{2}{4} \div \frac{2}{2} = \frac{1}{2}$$

Turn to page 70 for a review of simplifying fractions.

A. Add or subtract the following fractions. Express fractions in lowest terms. Change improper fractions to whole or mixed numbers.

1. $\frac{4}{5} - \frac{1}{5}$ $\frac{2}{4} + \frac{3}{4}$ $\frac{7}{10} - \frac{3}{10}$ $\frac{8}{9} + \frac{7}{9}$

2. $\frac{3}{8} - \frac{1}{8}$ $\frac{5}{2} + \frac{7}{2}$ $\frac{11}{12} - \frac{7}{12}$ $\frac{5}{6} + \frac{5}{6}$

B. Solve the following word problems. Pay careful attention to whether you should add or subtract.

3. Colin spent $\frac{3}{4}$ hour driving to the job interview and $\frac{1}{4}$ hour finding a place to park. How long did he spend driving and parking?

   (1) $\frac{1}{4}$ hour
   (2) $\frac{1}{2}$ hour
   (3) 1 hour

4. Art ran a sheet of wood $\frac{9}{16}$ inch thick through a sander that reduced the wood's thickness $\frac{1}{16}$ inch. What was the thickness of the wood after sanding?

   (1) $\frac{3}{8}$ inch
   (2) $\frac{1}{2}$ inch
   (3) $\frac{5}{8}$ inch

5. In a driving test, Marta's reaction time was $\frac{7}{100}$ second. Todd's time was $\frac{12}{100}$ second. How much faster was Marta's time than Todd's?

   (1) $\frac{1}{5}$ second
   (2) $\frac{19}{100}$ second
   (3) $\frac{1}{20}$ second

6. To dye eggs, Minako adds a few drops of food coloring to $\frac{5}{8}$ cup of hot water. She then adds $\frac{1}{8}$ cup of vinegar. How many cups of dye does this recipe make?

   (1) $\frac{1}{2}$ cup
   (2) $\frac{3}{4}$ cup
   (3) 1 cup

C. Adam, a computer technician, schedules service calls by the quarter hour. The chart below shows how many quarter hours are estimated for each job. Use the chart to answer the questions.

Job	Quarter Hours Scheduled
Hard Drive Installation	2
Sound Card Installation	1
CD-ROM Installation	3
Hard Drive Diagnostic	4
System Upgrade	5
RAM Upgrade	2

7. How much time will Adam schedule to install a sound card and CD-ROM?

8. Adam has $1\frac{1}{2}$ hours available. Does he have time to do a system upgrade and hard drive diagnostic?

9. **Multiple Solutions** Which jobs or combination of jobs would take *less* than 1 hour?

Finding Common Denominators

Remember that unlike fractions are fractions with different denominators. Before you add or subtract unlike fractions, you must give them the same denominator. To do this, write equivalent fractions.

You know that you can multiply both the numerator and denominator of a fraction by the same number *without changing the value of that fraction.*

Tip

Any number over itself is equal to 1. Multiplying $\frac{1}{3}$ by $\frac{2}{2}$ is the same as multiplying $\frac{1}{3}$ by 1.

Writing an Equivalent Fraction

Example: $\frac{3}{4} = \frac{\blacksquare}{8}$

Step 1
Figure out what amount the first denominator was multiplied by to get the second denominator.

$$\frac{3}{4} \times ? = \frac{\blacksquare}{8} \qquad 4 \times 2 = 8$$

Step 2
Multiply the first numerator by the same amount to get the second numerator.

$$\frac{3}{4} \times \frac{2}{2} = \frac{6}{8}$$

$\frac{3}{4}$ is equal to $\frac{6}{8}$.

A. **Write equivalent fractions. The first one in each row is started.**

1. $\overset{\times 2}{\underset{\times 2}{\frac{3}{8}}} = \frac{\blacksquare}{16}$ $\qquad \frac{1}{2} = \frac{\blacksquare}{12} \qquad \frac{2}{7} = \frac{\blacksquare}{21} \qquad \frac{5}{8} = \frac{\blacksquare}{40} \qquad \frac{5}{6} = \frac{\blacksquare}{18}$

2. $\overset{\times 4}{\underset{\times 4}{\frac{2}{9}}} = \frac{\blacksquare}{36}$ $\qquad \frac{3}{5} = \frac{\blacksquare}{50} \qquad \frac{1}{4} = \frac{\blacksquare}{48} \qquad \frac{2}{3} = \frac{\blacksquare}{33} \qquad \frac{7}{10} = \frac{\blacksquare}{100}$

3. $\overset{\times 5}{\underset{\times 5}{\frac{8}{11}}} = \frac{\blacksquare}{55}$ $\qquad \frac{3}{4} = \frac{\blacksquare}{20} \qquad \frac{4}{5} = \frac{\blacksquare}{100} \qquad \frac{4}{7} = \frac{\blacksquare}{28} \qquad \frac{5}{12} = \frac{\blacksquare}{60}$

Choosing a Common Denominator

You can use your knowledge of equivalent fractions to find a **common denominator** for two or more fractions.

Finding a Common Denominator

Example: What is a common denominator for $\frac{1}{3}$ and $\frac{3}{4}$?

Step 1
List the multiples of each denominator.

Step 2
Choose the lowest common multiple for both fractions.

Step 3
Write equivalent fractions using this new denominator.

$\frac{1}{3}$ multiples of 3:
3, 6, 9, 12, 18

$\frac{1}{3}$
3 (× 1)   6 (× 2)   9 (× 3)   ⑫ (× 4)   15 (× 5)

$\frac{1}{3} = \frac{4}{12}$ $\frac{3}{4} = \frac{9}{12}$

$\frac{3}{4}$ multiples of 4:
4, 8, 12, 16, 20

$\frac{3}{4}$
4 (× 1)   8 (× 2)   ⑫ (× 3)   16 (× 4)   20 (× 5)

The new denominator **12** is the lowest multiple both denominators have in common.

B. **For each number in the pairs below, list the first five multiples (or more, if necessary). Then circle the common multiple. The first one is done for you.**

4. 3: _3, 6, 9, 12, ⑮_   2: _____   4: _____

   5: _5, 10, ⑮, 20, 25_   5: _____   6: _____

5. 4: _____   6: _____   9: _____

   10: _____   9: _____   12: _____

C. **Apply your fraction skills to this problem about time.**

6. **Draw** Fran spent $\frac{5}{8}$ hour working on her math assignment. Joe spent $\frac{3}{4}$ hour. The circles at the right each represent 1 hour. Shade sections to show what part of an hour each person worked. Who worked longer? How can you use your knowledge of common denominators and equivalent fractions to check your answer?

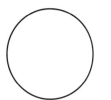

Adding and Subtracting Unlike Fractions

Before you can add or subtract unlike fractions, you must turn them into equivalent like fractions.

Adding Unlike Fractions

Example: Carol has a bag with $\frac{2}{3}$ cup of sugar in it. She also has $\frac{1}{2}$ cup of sugar in a sugar bowl. How much sugar does Carol have in all?

Step 1
Choose a common denominator by using multiples.

$$\frac{2}{3} \quad 3, \circled{6}$$
$$\quad\quad (\times 2)$$
$$+ \frac{1}{2} \quad 2, 4, \circled{6}$$
$$\quad\quad (\times 3)$$

Step 2
Write equivalent fractions using the new denominator.

$$\frac{2}{3} = \frac{4}{6}$$

$$\frac{1}{2} = \frac{3}{6}$$

Step 3
Add the like fractions.

$$\frac{4}{6}$$
$$+ \frac{3}{6}$$
$$\overline{\frac{7}{6}}$$

Step 4
Simplify and write a mixed number as necessary.

$$\frac{7}{6} = 6\overline{)7}^{\,1\,R1} = 1\frac{1}{6}$$
$$\quad\quad \underline{-\,6}$$
$$\quad\quad\quad 1$$

Your answer is $1\frac{1}{6}$ **cups.**

Subtracting Unlike Fractions

Example: What is $\frac{2}{3} - \frac{1}{6}$?

Step 1
Choose a common denominator by using multiples.

$$\frac{2}{3} \quad 3, \circled{6}$$
$$\quad\quad (\times 2)$$
$$- \frac{1}{6} \quad 6$$

Step 2
Write equivalent fractions using the new denominator.

$$\frac{2}{3} = \frac{4}{6}$$

$$\frac{1}{6} = \frac{1}{6}$$

Step 3
Subtract the like fractions.

$$\frac{4}{6}$$
$$- \frac{1}{6}$$
$$\overline{\frac{3}{6}}$$

Step 4
Simplify.

$$\frac{3}{6} \div \frac{3}{3} = \frac{1}{2}$$

Your answer is $\frac{1}{2}$.

A. Add or subtract the following unlike fractions. Simplify and write as a
    mixed number as necessary.

1. $\frac{4}{9} + \frac{1}{3}$ $\qquad\qquad$ $\frac{7}{8} + \frac{3}{4}$ $\qquad\qquad$ $\frac{5}{6} + \frac{1}{9}$ $\qquad\qquad$ $\frac{1}{2} + \frac{7}{10}$

2. $\frac{3}{4} - \frac{2}{3}$ $\qquad\qquad$ $\frac{8}{9} - \frac{5}{6}$ $\qquad\qquad$ $\frac{7}{10} - \frac{1}{4}$ $\qquad\qquad$ $\frac{2}{3} - \frac{3}{5}$

B. Solve the following word problems. Pay careful attention to whether you
    should add or subtract.

3. Kimiko has a piece of wood $\frac{5}{16}$ inch long.
    How much wood will she have left if she
    trims off $\frac{1}{16}$ inch?
    (1) $\frac{1}{16}$ inch
    (2) $\frac{1}{4}$ inch
    (3) $\frac{1}{8}$ inch

4. Pedro had part of a gallon of window
    washer fluid in his garage. His wife and son
    each added washer fluid to their cars. At the
    end of the week, he had $\frac{1}{4}$ gallon left. If there
    was $\frac{9}{10}$ gallon at the beginning of the week,
    how much fluid did his family use?
    (1) $\frac{4}{5}$ gallon
    (2) $\frac{13}{20}$ gallon
    (3) $\frac{2}{5}$ gallon

5. Lindsey bought $\frac{3}{8}$ pound of shredded
    cheese. She already had $\frac{1}{3}$ pound at home.
    How much shredded cheese does Lindsey
    have altogether?
    (1) $\frac{4}{11}$ pound
    (2) $\frac{1}{6}$ pound
    (3) $\frac{17}{24}$ pound

6. It normally takes Boris $\frac{3}{4}$ of an hour to get
    to work by bus. Today he decided to drive
    his car instead. If Boris got to work in $\frac{1}{2}$ of
    an hour, how much sooner did he get to
    work?
    (1) $\frac{1}{4}$ hour
    (2) $\frac{1}{8}$ hour
    (3) $\frac{2}{3}$ hour

C. Ricardo usually buys a pizza cut into 8 equal pieces. This time he orders the
    same size pizza cut into 6 pieces.

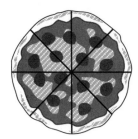 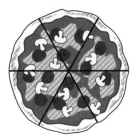

7. How much more pizza will Ricardo eat if
    he has 3 pieces of the 6-slice pizza instead
    of 3 pieces of the 8-slice pizza?

8. **Multiple Answers** If Ricardo buys both
    pizzas, what combinations of slices would
    add up to less than $\frac{1}{3}$ of a pizza?

Working with Distances

Calculating distances often requires the addition and subtraction of mixed numbers.

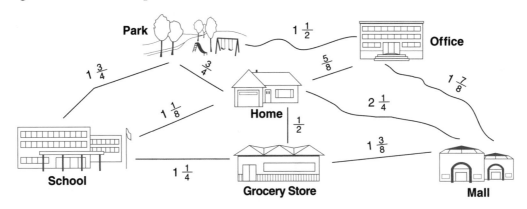

Distances are shown in miles.

Adding Distances

Example: According to the map above, how many miles would you walk if you went from home to the mall to the grocery store and home again?

Step 1
Change the fractions into like fractions.

$$2\tfrac{1}{4} = 2\tfrac{2}{8}$$
$$1\tfrac{3}{8} = 1\tfrac{3}{8}$$
$$+ \quad \tfrac{1}{2} = \quad \tfrac{4}{8}$$

Step 2
Add the like fractions. Simplify and turn an improper fraction into a mixed number.

$$\tfrac{2}{8}$$
$$\tfrac{3}{8}$$
$$+ \quad \tfrac{4}{8}$$
$$\overline{\quad \tfrac{9}{8} = 1\tfrac{1}{8}}$$

Step 3
Add the whole numbers. Then add the mixed number to the whole-number sum.

$$2\tfrac{2}{8}$$
$$1\tfrac{3}{8}$$
$$+ \quad \tfrac{4}{8}$$
$$\overline{3 + 1\tfrac{1}{8} = \mathbf{4\tfrac{1}{8}\ miles}}$$

Subtracting Distances

Example: How much farther is it from the mall to home than the mall to the grocery store?

Step 1
Change the fractions into like fractions.

$$2\tfrac{1}{4} = 2\tfrac{2}{8}$$
$$- 1\tfrac{3}{8} = 1\tfrac{3}{8}$$

Step 2
Subtract the fractions. If necessary, regroup (borrow) 1 from the whole number. Write the 1 as a like fraction and add it to the first number.

$$2\tfrac{2}{8} = 1\tfrac{2}{8} + \tfrac{8}{8}$$
$$- 1\tfrac{3}{8} = 1\tfrac{3}{8}$$

Step 3
Subtract. Simplify, if necessary.

$$1\tfrac{10}{8}$$
$$- 1\tfrac{3}{8}$$
$$\overline{\quad \tfrac{7}{8}\ \mathbf{mile}}$$

A. Add or subtract the following numbers. Express fractions in lowest terms. Change improper fractions to mixed numbers.

1. $1\frac{1}{3} + 4\frac{5}{6}$ $2\frac{1}{8} + \frac{3}{4} + \frac{1}{3}$ $5\frac{1}{5} + 2\frac{9}{10}$ $2\frac{1}{6} + 1\frac{4}{9} + 3\frac{2}{3}$

2. $5\frac{2}{3} - 3\frac{1}{4}$ $4\frac{3}{8} - 1\frac{3}{4}$ $7\frac{5}{6} - 2\frac{1}{5}$ $3\frac{1}{7} - 1\frac{2}{3}$

B. Solve the following word problems using the map on page 80. Pay careful attention to whether you should add or subtract.

3. How many miles will Eduardo walk if he goes from home to school, to the park, and then home again?

4. Gabriella wants to go to the mall from school. How much shorter would the route be if she went by the grocery store rather than going by her home?

5. How much farther will Maria walk if she goes from the grocery store to the school and to the park than if she goes from the grocery store to home and to the park?

6. Which route should Mahmoud take to go from the mall to the park if he wants to walk the shortest distance?

Making Connections: Distance and Exercise

Walking is a very popular form of exercise. Steve discussed an exercise program with his doctor. He needs to walk 2 miles a day for one month. Use the map below and answer the questions to help Steve prepare his daily walks.

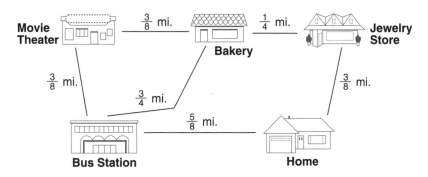

1. Describe a path Steve could take to walk 2 miles starting from his home and ending by returning to his home.

2. At the beginning of the fourth month, Steve should be walking *at least* $2\frac{1}{2}$ miles a day. Describe a path for Steve to walk starting from and returning to his home.

Multiplying Fractions

When you multiply a fraction by a fraction, the answer will be smaller than the fractions you started with. (*Note:* This is not true for improper fractions.)

Multipying Fractions

Example 1: Sylvia had $\frac{3}{4}$ pound of candy. She wanted to give each of her 3 grandchildren $\frac{1}{3}$ of the candy. How much candy should Sylvia give each grandchild?

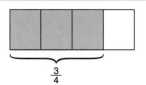

$$\frac{3}{4}$$

Step 1
Multiply $\frac{3}{4}$ by $\frac{1}{3}$. Look for ways to cancel (or factor out) a multiple common to the numerator and the denominator. This makes the problem simpler.

$\frac{3}{4} \times \frac{1}{3} =$

Step 2
Cancel the numerator of the first fraction and the denominator of the second fraction.

$\overset{1}{\cancel{\frac{3}{4}}} \times \frac{1}{\underset{1}{\cancel{3}}} =$

Step 3
Multiply across using the new numerators and denominators.

$\frac{1}{4} \times \frac{1}{1} = \frac{1}{4}$ **pound**

Example 2: Multiply $\frac{6}{7}$ by $\frac{7}{9}$.

Step 1
Look for ways to cancel. Notice that canceling may include both numerators and denominators.

$\frac{6}{7} \times \frac{7}{9}$

Step 2
Cancel the numerator of the first fraction and the denominator of the second fraction.

$\overset{2}{\cancel{\frac{6}{7}}} \times \frac{7}{\underset{3}{\cancel{9}}}$

Step 3
Cancel the numerator of the second fraction and the denominator of the first fraction.

$\underset{1}{\overset{2}{\cancel{\frac{6}{7}}}} \times \overset{1}{\cancel{\frac{7}{9}}}_{3}$

Step 4
Multiply across using the new numerators and denominators.

$\frac{2}{1} \times \frac{1}{3} = \frac{2}{3}$

If you did not cancel the fractions before multiplying, you would still get the same answer. However, the answer would have a larger numerator and denominator, and you would have to simplify your answer after multiplying. Therefore, it is much easier to cancel first.

A. **Multiply the following fractions. Use canceling to make the work easier, or simplify your answers.**

1. $\overset{1}{\cancel{\frac{2}{5}}} \times \frac{1}{\underset{1}{\cancel{2}}}$ $\frac{7}{8} \times \frac{2}{3}$ $\frac{4}{5} \times \frac{3}{4}$ $\frac{1}{6} \times \frac{5}{8}$

2. $\underset{4}{\overset{1}{\cancel{\frac{3}{8}}}} \times \overset{1}{\cancel{\frac{2}{3}}}_{1}$ $\frac{9}{10} \times \frac{2}{3}$ $\frac{4}{9} \times \frac{3}{8}$ $\frac{6}{7} \times \frac{1}{3}$

B. Choose the correct answer for each problem below.

3. Mustafa had a section of paneling $\frac{2}{3}$ yard long. He had to use $\frac{1}{2}$ of the section to repair a wall. How much of the paneling did Mustafa use?

 (1) $\frac{1}{2}$ yard

 (2) $\frac{2}{3}$ yard

 (3) $\frac{1}{3}$ yard

4. Tony has $\frac{7}{8}$ pound of hamburger meat in the freezer. If he uses $\frac{2}{3}$ of the meat to make meatballs, how much of the meat will Tony have used?

 (1) $\frac{7}{12}$ pound

 (2) $\frac{3}{4}$ pound

 (3) $\frac{5}{7}$ pound

5. Margaret's recipe called for her to grate $\frac{3}{4}$ cup of cheese. The next step was to set aside $\frac{1}{2}$ of the cheese. How much of the cheese is to be set aside?

 (1) $\frac{3}{2}$ cup

 (2) $\frac{3}{4}$ cup

 (3) $\frac{3}{8}$ cup

6. Five office workers decide to split $\frac{15}{16}$ fluid ounce of perfume evenly. If they each get $\frac{1}{5}$ of the perfume, how much will each of them get?

 (1) $\frac{3}{8}$ fluid ounce

 (2) $\frac{1}{5}$ fluid ounce

 (3) $\frac{3}{16}$ fluid ounce

C. Min is taking a computer class. The instructor separated the class in half. Then one of the halves was to break into 3 equal groups.

7. What fraction of the entire class is in each of the 3 smaller groups?

8. If there are 24 students in the class, how many are in each of the 3 smaller groups?

9. The instructor estimates that $\frac{1}{4}$ of the students in the smaller groups will miss a class while at least $\frac{1}{2}$ of the students in the larger group will miss a class. According to the instructor's estimates

 a. How many students would be absent in the smaller groups?

 b. How many students in the larger group would be absent?

10. **Discuss** The result of multiplying a fraction by a fraction is always less than the fractions you start with. (*Remember:* This doesn't apply to improper fractions.) Why is this so? Try multiplying a large fraction by a very small fraction to see if the result is less than either fraction you multiplied. Compare your results with other students' results.

Dividing Fractions

To divide a fraction, invert the divisor (the number you are dividing by) and then multiply the fractions.

You can simplify the multiplication if you can cancel the fractions first.

When you divide a fraction by a fraction, the answer will be larger than the fractions you started with. (*Note:* This doesn't apply to improper fractions.) This is because you are finding how many of the second fraction there are in the fraction being divided up.

Dividing by a Fraction

Example: Angela has to give her cat $\frac{1}{4}$ of a pill twice a day. If she has $\frac{1}{2}$ of a pill left, how many more times can she give her cat its medicine?

Step 1
Invert the divisor (the number you are dividing *by*).

$\frac{1}{2} \div \frac{1}{4}$

$\frac{1}{4}$ inverts to $\frac{4}{1}$

Step 2
Change the ÷ into a ×.
Use canceling, if possible.

$\frac{1}{\underset{1}{2}} \times \frac{\overset{2}{4}}{1}$

Step 3
Multiply across using the new numerators and denominators.

$\frac{1}{1} \times \frac{2}{1} = 2$

Angela can give her cat $\frac{1}{4}$ of a pill **2** more times.

When you divide a fraction by a whole number, the answer is a smaller number than the fraction.

Dividing by a Whole Number

Example: The Morrison family has $\frac{3}{4}$ of a pie left over. If there are 6 people in the family, what fraction of the original pie will each person get?

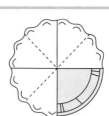

Step 1
Write the whole number as a fraction with a denominator of 1.
Then invert the fraction.

$\frac{3}{4} \div \frac{6}{1}$

$\frac{6}{1}$ inverts to $\frac{1}{6}$

Step 2
Change the ÷ into a ×.
Use canceling, if possible.

$\frac{\overset{1}{3}}{4} \times \frac{1}{\underset{2}{6}}$

Step 3
Multiply across using the new numerators and denominators.

$\frac{1}{4} \times \frac{1}{2} = \frac{1}{8}$

Each person will get $\frac{1}{8}$ of the pie.

A. Divide the following fractions and whole numbers.

1. $\frac{3}{4} \div \frac{4}{5}$ $\qquad$ $\frac{7}{8} \div \frac{7}{10}$ $\qquad$ $\frac{1}{2} \div \frac{3}{4}$ $\qquad$ $\frac{8}{9} \div \frac{2}{3}$

2. $\frac{1}{4} \div 3$ $\qquad$ $\frac{5}{6} \div 15$ $\qquad$ $8 \div \frac{4}{9}$ $\qquad$ $\frac{2}{5} \div \frac{3}{7}$

B. Predict whether the results of these divisions will be larger or smaller than the fraction being divided.

3. $\frac{5}{9} \div \frac{2}{11}$ $\qquad$ $\frac{4}{7} \div 5$ $\qquad$ $\frac{3}{8} \div \frac{6}{1}$ $\qquad$ $\frac{8}{13} \div \frac{1}{2}$

Making Connections: Estimating with Fractions

To estimate when you are multiplying and dividing fractions, round only to $\frac{1}{2}$ or 1 if the fraction is close to those numbers. Small fractions should be left as they are.

Example 1: Estimate an answer to $\frac{8}{9} \times \frac{3}{8}$.

Round: $\frac{8}{9}$ to 1 and $\frac{3}{8}$ to $\frac{1}{2}$

Multiply: $1 \times \frac{1}{2} = \frac{1}{2}$

Check: $\frac{8}{9} \times \frac{3}{8} = \frac{1}{3}$

The estimate, **$\frac{1}{2}$,** is close to the exact answer, $\frac{1}{3}$.

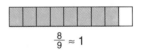

$\frac{8}{9} \approx 1$

$\frac{3}{8} \approx \frac{1}{2}$

Example 2: Estimate an answer to $\frac{5}{8} \div \frac{1}{16}$.

Round: $\frac{5}{8}$ to $\frac{1}{2}$; $\frac{1}{16}$ is too small to round to $\frac{1}{2}$ so leave it as is.

Divide: $\frac{1}{2} \div \frac{1}{16} = \frac{1}{2} \times \frac{16}{1} = \frac{8}{1} = 8$

Check: $\frac{5}{8} \div \frac{1}{16} = \frac{5}{8} \times \frac{16}{1} = \frac{10}{1} = 10$

The estimate, **8,** is close to the exact answer, 10.

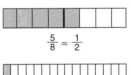

$\frac{5}{8} \approx \frac{1}{2}$

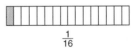

$\frac{1}{16}$

Estimate Use estimation to solve these multiplication and division problems. Then find the exact answer and compare it to your estimate.

1. $\frac{3}{5} \times \frac{6}{7}$ $\qquad$ 2. $\frac{7}{8} \times \frac{1}{3}$ $\qquad$ 3. $\frac{4}{5} \div \frac{3}{7}$ $\qquad$ 4. $\frac{3}{4} \div 9$

Dividing Fractions with Mixed Numbers

Dividing with mixed numbers is the same as dividing with fractions or whole numbers—once you change the mixed numbers into improper fractions.

Dividing Mixed Numbers

Example: Bill used $2\frac{1}{2}$ gallons of paint to cover one side of a warehouse. Each side is the same size. How many sides could he paint if he had $6\frac{1}{4}$ gallons of paint?

Step 1
Change mixed numbers to fractions.

$6\frac{1}{4} \div 2\frac{1}{2}$

$\frac{25}{4} \div \frac{5}{2}$

Step 2
Invert the divisor and change the $\div$ to a $\times$.

$\frac{25}{4} \times \frac{2}{5}$

Step 3
Cancel if possible and multiply.

$\frac{\overset{5}{\cancel{25}}}{\underset{2}{\cancel{4}}} \times \frac{\overset{1}{\cancel{2}}}{\underset{1}{\cancel{5}}}$

Step 4
Change improper fractions to mixed numbers.

$\frac{5}{2} = 2\frac{1}{2}$

Bill could paint **$2\frac{1}{2}$** sides with $6\frac{1}{4}$ gallons of paint.

A. **Divide the following amounts. Change the mixed numbers into improper fractions before dividing.**

1. $1\frac{1}{2} \div \frac{3}{4}$ $\qquad$ $2\frac{1}{8} \div \frac{1}{4}$ $\qquad$ $4\frac{3}{5} \div \frac{5}{8}$ $\qquad$ $\frac{2}{3} \div 1\frac{3}{4}$

2. $\frac{5}{8} \div 2\frac{1}{2}$ $\qquad$ $5 \div 3\frac{1}{3}$ $\qquad$ $6\frac{2}{3} \div 2\frac{1}{2}$ $\qquad$ $4\frac{3}{8} \div 5$

3. $1\frac{7}{8} \div 3\frac{3}{4}$ $\qquad$ $\frac{1}{4} \div 5\frac{1}{3}$ $\qquad$ $3\frac{1}{2} \div 2\frac{1}{3}$ $\qquad$ $8 \div 1\frac{1}{5}$

4. $2\frac{1}{5} \div 2\frac{1}{2}$ $\qquad$ $5\frac{5}{8} \div \frac{3}{4}$ $\qquad$ $6\frac{1}{6} \div 2$ $\qquad$ $3\frac{1}{7} \div \frac{11}{14}$

B. Solve the following problems.

5. Kristen bought a $12\frac{1}{4}$-fluid-ounce bottle of apple juice. How many glasses of juice will she have if she pours 4 fluid ounces into each glass?

   (1) $5\frac{1}{4}$ glasses

   (2) $3\frac{1}{16}$ glasses

   (3) 49 glasses

6. After carving the Thanksgiving turkey, the Peterson family had $10\frac{2}{3}$ pounds of meat. How many $\frac{1}{3}$-pound servings did they have?

   (1) 32 servings

   (2) $10\frac{2}{3}$ servings

   (3) $24\frac{1}{3}$ servings

7. Joe is wrapping holiday presents. He has a 24-foot roll of wrapping paper. How many presents can Joe wrap if each present takes $1\frac{1}{2}$ feet of paper?

   (1) $16\frac{3}{12}$ presents

   (2) $13\frac{1}{2}$ presents

   (3) 16 presents

8. A tree in Pablo's backyard has been growing $\frac{1}{2}$ foot each year. Assuming the tree grows at the same rate, how many years will it take for the tree to grow $6\frac{1}{2}$ feet?

   (1) $3\frac{1}{4}$ years

   (2) 13 years

   (3) $4\frac{1}{2}$ years

Making Connections: Estimating and Mixed Numbers

When estimating division with mixed numbers, round the mixed numbers to the nearest whole number. When you have two mixed numbers, you often get a closer estimate if you round only one of them.

Example: Estimate an answer to $6\frac{3}{4} \div 1\frac{7}{8}$.

$6\frac{3}{4} \approx 7$ and $1\frac{7}{8} \approx 2$

$7 \div 2 = 3\frac{1}{2}$

Check: $6\frac{3}{4} = \frac{27}{4}$ $1\frac{7}{8} = \frac{15}{8}$

$\frac{27}{4} \div \frac{15}{8} = \frac{27}{4} \times \frac{8}{15} = \frac{18}{5} = 3\frac{3}{5}$

The answer, $3\frac{3}{5}$, is close to the estimate of $3\frac{1}{2}$.

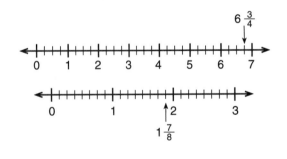

Estimate Use estimation to solve these division problems. Then find the exact answers to see how close your estimates were.

1. $4\frac{1}{3} \div 2\frac{1}{8}$ **2.** $6 \div \frac{6}{7}$ **3.** $3\frac{3}{4} \div 1\frac{5}{8}$ **4.** $5\frac{1}{2} \div \frac{3}{4}$

Mixed Review

A. Choose each answer from the choices given.

1. The mixed number $1\frac{1}{3}$ is the same as the fraction
   - (1) $\frac{11}{3}$
   - (2) $\frac{4}{3}$
   - (3) $\frac{5}{3}$

2. Which fraction is larger than $\frac{2}{3}$?
   - (1) $\frac{8}{12}$
   - (2) $\frac{2}{4}$
   - (3) $\frac{3}{4}$

3. Which metric measurement is the same as 58 millimeters?
   - (1) 58 centimeters
   - (2) 5 millimeters 8 centimeters
   - (3) 5.8 centimeters

4. Which fraction is equivalent to $\frac{3}{4}$?
   - (1) $\frac{9}{12}$
   - (2) $\frac{8}{12}$
   - (3) $\frac{3}{12}$

5. Which is a common denominator for fifths and thirds?
   - (1) eighths
   - (2) fifteenths
   - (3) fifths

6. Which is a good estimate for the multiplication of $\frac{7}{8}$ and $\frac{5}{8}$?
   - (1) $1 \times \frac{1}{2}$
   - (2) 1×1
   - (3) $\frac{1}{2} \times \frac{1}{2}$

B. Solve the following problems. Simplify answers to lowest terms and change improper fractions to mixed numbers.

7. $\frac{7}{12} + \frac{1}{12}$ $\qquad$ $\frac{3}{4} + \frac{3}{4}$ $\qquad$ $\frac{5}{8} + \frac{1}{2}$ $\qquad$ $\frac{2}{9} + \frac{5}{6}$

8. $\frac{7}{8} - \frac{1}{8}$ $\qquad$ $1\frac{1}{3} - \frac{2}{3}$ $\qquad$ $\frac{5}{6} - \frac{3}{4}$ $\qquad$ $3\frac{1}{2} - \frac{7}{9}$

9. $\frac{1}{3} \times \frac{1}{3}$ $\qquad$ $\frac{3}{4} \times \frac{2}{3}$ $\qquad$ $\frac{7}{12} \times \frac{4}{5}$ $\qquad$ $\frac{9}{10} \times \frac{4}{21}$

10. $\frac{1}{2} \div \frac{3}{4}$ $\qquad$ $\frac{5}{6} \div \frac{3}{7}$ $\qquad$ $\frac{5}{8} \div 1\frac{2}{3}$ $\qquad$ $4\frac{1}{2} \div \frac{3}{8}$

C. Solve the following problems. Put answers in lowest terms and change improper fractions to mixed numbers.

11. Marcia has two bags of fruit. One bag weighs $3\frac{1}{3}$ pounds. The other bag weighs $2\frac{1}{2}$ pounds. How much do the two bags weigh together?

12. Each year the Williams family measures and records their son's height. The illustration below shows his height last year and this year. How much did their son grow since last year?

— $60\frac{1}{2}$ in. (this year)

— $56\frac{3}{4}$ in. (last year)

13. Bart has two bottles of juice. One bottle has $24\frac{1}{2}$ fluid ounces of juice. The other bottle holds $34\frac{1}{4}$ fluid ounces of juice, but $\frac{1}{4}$ of it has been drunk. Estimate which bottle has more juice in it.

14. Jaime bought $5\frac{1}{3}$ pounds of holiday cookies. He wants to arrange 4 cookie trays. How much will each tray hold if he divides the cookies evenly?

15. $\frac{2}{3}$ of a pumpkin pie is left over from a Thanksgiving dinner. If Barbara cuts off $\frac{1}{6}$ of the remaining pie, what fraction of the whole pie will she take?

16. Mia is preparing pots for her tomatoes. She has $3\frac{1}{4}$ pounds of peat moss. How many pots can she prepare if each pot requires $\frac{1}{3}$ pound of peat moss?

D. Use the map to solve problems 17–20.

17. What is the total distance from Kim's to the restaurant to the beauty salon to the municipal building and back to Kim's?

18. How much farther is it from Kim's to the restaurant than it is from the restaurant to the beauty salon?

19. Craig got a flat tire $\frac{2}{3}$ of the way from the municipal building to Kim's. How far had Craig ridden from the municipal building?

20. **Multiple Solutions** The distance from the restaurant to the municipal building is $1\frac{1}{4}$ miles, if you ride straight there. How much distance would be saved by not going to the beauty salon first? List two ways that you could find the answer.

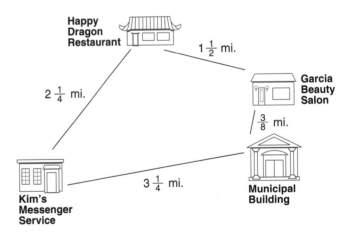

Relating Fractions and Ratios

A ratio compares one number to another. A fraction is a kind of ratio that compares a part to a whole.

Example: Lily has 16 coins. 5 of the coins are pennies, 10 are dimes, and 1 is a quarter. The fraction $\frac{5}{16}$ expresses the number of pennies (part) compared to the total number of coins (whole). It can also be considered a ratio.

$$\frac{5}{16} \begin{array}{l} \text{— pennies} \\ \text{— total coins} \end{array}$$

The comparison of dimes to pennies also is a ratio. Like all ratios, it can be written in three different forms.

Fraction Form	With a Colon	Using *to*
$\frac{10}{5}$	10:5	10 to 5

▶ Ratios can be simplified to lowest terms, just like fractions. As with fractions, divide both numbers in the ratio by the same whole number.

Put in lowest terms:

$\frac{10}{5} = \frac{2}{1}$ *or* 2:1 *or* 2 to 1

Tip

When writing a ratio in fraction form, always keep a denominator. For example, a ratio of 2 to 1 would be written as $\frac{2}{1}$.

A. Write ratios based on the items in the illustration. Use any of the three forms you learned on page 90. Simplify to lowest terms.

1. cups of water to tablespoons of flour _____

2. eggs to sticks of butter _____

3. cups of water to eggs _____

4. sticks of butter to tablespoons of flour _____

5. tablespoons of flour to eggs _____

6. eggs to tablespoons of flour _____

B. Use the figures below to make up your own ratios. Be sure you include labels as shown in problem 7, which is done as an example. Simplify to lowest terms.

7. $\frac{circles}{wavy\ lines} = \frac{6}{5}$

8.

9.

10.

11.

12.

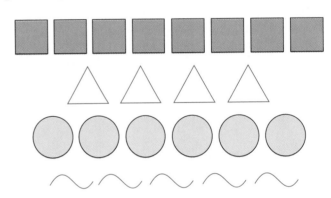

C. In an adult education class, 6 students are Hispanic, 8 students are Asian, 5 students are African American, and the other 11 students are Caucasian.

13. What is the ratio of Caucasian students to the total number of students?

15. What is the ratio of Asian students to African American students?

14. Which ratios of students would change if 2 Hispanic students were added to the class?

16. **Draw** Use symbols or figures to illustrate the ratio of Hispanic students to Asian students in the class.

Writing Ratios

The relationship of one number to another can be written as a ratio. When writing a ratio in fraction form, the first number becomes the numerator and the second number becomes the denominator. Ratios can be simplified to lowest terms.

Writing a Ratio

Example: During a 30-minute television show, there were 8 minutes of commercials. Write a ratio of commercial time to the total length of time allowed for the show.

Step 1
Use the first number in the ratio as a numerator. Make the second number the denominator.

$$\frac{8 \text{ minutes of commercials}}{30 \text{ minutes of show}}$$

Step 2
Simplify if necessary.

$$\frac{8}{30} = \frac{4}{15}$$

The ratio is $\frac{4}{15}$, **4:15**, or **4 to 15** minutes of commercials to the total length of the television show.

> The order in which you place the numbers is important. $\frac{4}{15}$ is not the same as $\frac{15}{4}$. Labeling the fractions helps to keep the ratios in the correct order.

When you compare measurements that are in different units, convert one measurement so that both measurements have the same units.

Converting Units to Write a Ratio

Example: Carl is drawing on posterboard measuring 2 feet wide and 30 inches long. What is the ratio of width to length of the posterboard?

Step 1
Convert to the same unit of measurement.

2 feet = 24 inches

Step 2
Set up a ratio in fraction form.

$$\frac{24 \text{ inches wide}}{30 \text{ inches long}}$$

Step 3
Simplify if necessary.

$$\frac{4}{5}$$

The ratio is $\frac{4}{5}$, **4:5**, or **4 to 5** inches of width to length.

A. Write ratios for each of the following situations. Be sure to express ratios in lowest terms.

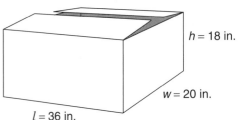

h = 18 in.

w = 20 in.

l = 36 in.

1. Write the ratio of the width (*w*) to the height (*h*) of the box.

2. Write the ratio of the width (*w*) to the length (*l*) of the box.

3. Write the ratio of the length (*l*) to the height (*h*) of the box.

B. Use the information in the chart below to solve problems 4–8. Express ratios in lowest terms.

SOFTBALL GAME STATISTICS

Singles	8
Doubles	4
Triples	2
Home Runs	2
Total Hits	**16**

4. What is the ratio of singles to total hits?

5. What is the ratio of home runs to triples?

6. What is the ratio of singles and doubles to total hits?

7. What is the ratio of singles to doubles?

8. What is the ratio of triples to total hits?

C. Use the table below to solve problems 9–11. Express ratios in lowest terms.

Job	All Employees		Minority Employees	
	Male	Female	Male	Female
Manager	6	3	1	1
Sales	5	4	2	2
Clerical	1	8	—	6
Customer Service	3	15	2	10

Some institutions ask companies with which they want to do business to provide personnel statistics. The table above is a sample of such information.

9. What is the ratio of female to male managers? (*Hint:* Look under *All Employees.*)

10. What is the ratio of managers to total employees? (*Hint:* Use the numbers under *All Employees.*)

11. What is the ratio of minority women to all female employees in customer service?

Writing Proportions

A **proportion** is made up of two equal ratios. When you work with equivalent fractions, you are working with a proportion.

▶ Two equal ratios make a proportion. Two ratios are equal if their **cross products** are equal.

$\frac{1}{2} \times\!\!=\!\!\times \frac{3}{6}$

Cross products: $1 \times 6 = \mathbf{6}$ and $3 \times 2 = \mathbf{6}$

If one number is missing in a proportion, you can use cross products and equations to find out what it is.

Solving Proportions

Example: $\frac{3}{4} = \frac{9}{?}$

Step 1
Use a variable for the number you don't know.

$\frac{3}{4} = \frac{9}{p}$

Step 2
Multiply to get the cross products and write an equation.

$3 \times p = 9 \times 4$

Step 3
Solve the equation.

$3 \times p = 9 \times 4$
$3 \times p = 36$
$\frac{3 \times p}{3} = \frac{36}{3}$
$p = 12$

Since you are dividing by 3 to get the p alone, you can write the third line above as $p = 36 \div 3$

Step 4
Check your proportion by using cross products.

$\frac{3}{4} = \frac{9}{\mathbf{12}}$
$3 \times 12 = 4 \times 9$
$36 = 36$

A. Use cross products to decide if the following ratios are equal.

1. $\frac{2}{3} = \frac{8}{12}$ $\frac{2}{6} = \frac{1}{4}$ $\frac{3}{4} = \frac{6}{8}$ $\frac{3}{9} = \frac{1}{3}$

2. $\frac{4}{7} = \frac{8}{14}$ $\frac{8}{18} = \frac{2}{9}$ $\frac{1}{2} = \frac{8}{12}$ $\frac{6}{15} = \frac{2}{5}$

3. $\frac{4}{5} = \frac{12}{15}$ $\frac{12}{18} = \frac{1}{2}$ $\frac{10}{15} = \frac{2}{3}$ $\frac{8}{9} = \frac{24}{27}$

B. Use cross products and equations to find the missing number in each proportion. You may use a calculator.

4. $\dfrac{3}{4} = \dfrac{6}{a}$ $\dfrac{2}{3} = \dfrac{b}{9}$ $\dfrac{4}{c} = \dfrac{16}{20}$ $\dfrac{d}{3} = \dfrac{10}{15}$

5. $\dfrac{5}{7} = \dfrac{15}{e}$ $\dfrac{3}{8} = \dfrac{f}{24}$ $\dfrac{1}{g} = \dfrac{8}{16}$ $\dfrac{h}{12} = \dfrac{1}{6}$

6. $\dfrac{16}{2} = \dfrac{8}{i}$ $\dfrac{3}{10} = \dfrac{j}{100}$ $\dfrac{5}{k} = \dfrac{25}{5}$ $\dfrac{l}{7} = \dfrac{18}{21}$

7. $\dfrac{7}{8} = \dfrac{49}{m}$ $\dfrac{3}{5} = \dfrac{n}{50}$ $\dfrac{5}{p} = \dfrac{20}{24}$ $\dfrac{q}{8} = \dfrac{60}{96}$

Making Connections: Making a Table

As you know, it can be difficult to decide what to do with all the numbers in a math problem. By making a chart or table, you can often tell whether a proportion can be used to solve a problem.

Example: In a recent football game, the Giants were given 4 penalties in 12 minutes. How many penalties would they have received in 48 minutes if they were penalized at the same rate?

Can a proportion be used to solve this problem? Make a table to find out. Use the labels that go with the numbers to set it up.

Penalties	4	x
Minutes	12	48

$$\dfrac{4 \text{ penalties}}{12 \text{ minutes}} = \dfrac{x \text{ penalties}}{48 \text{ minutes}}$$

$$4 \times 48 = 12 \times x$$

When you can make a table that compares two units (such as *penalties* and *minutes*), and you can fill in 3 of 4 values, you can write a proportion.

$$\dfrac{4 \times 48}{12} = \dfrac{12 \times x}{12}$$

$$\dfrac{192}{12} = x$$

$$\mathbf{16} = x$$

Make a table to solve each of these proportion problems.

1. Hariko used 4 cups of flour to make 20 muffins. How many muffins would 16 cups make?

2. Barry hit 5 home runs in 12 games. At this rate, how many home runs would he hit in 60 games?

Solving Problems with Proportions

You can use proportions to solve many kinds of problems.

▶ To solve using proportion, be sure that the two ratios have corresponding units.

Using Proportion

Example: At Bright's Movie Theater, 3 tickets cost $15. How much would 4 tickets cost?

Step 1
Write a ratio with two numbers from the problem. Include labels.

$$\frac{3 \text{ tickets}}{\$15}$$

Step 2
Write a proportion, using a variable for the number you do not know.

$$\frac{3 \text{ tickets}}{\$15} = \frac{4 \text{ tickets}}{\$p}$$

Step 3
Multiply to get the cross products and find the unknown value.

$$3 \times p = 15 \times 4$$
$$3 \times p = 60$$
$$p = 60 \div 3$$
$$\boldsymbol{p = \$20}$$

The cost of 4 tickets is **$20.**

Another way to set up a proportion for this problem is this: $\dfrac{\$15}{3 \text{ tickets}} = \dfrac{\$p}{4 \text{ tickets}}$

It doesn't matter which number goes on top or bottom in the first ratio. What *does* matter is that both ratios are set up in the same order with corresponding units.

For example, in the problem above, writing the two ratios as

$$\frac{\$15}{3 \text{ tickets}} = \frac{4 \text{ tickets}}{\$20}$$

would *not* be a true proportion because they do not compare the same units in their numerators or the same units in their denominators.

Another way to express the same proportion is

$$\frac{\$15}{\$p} = \frac{3 \text{ tickets}}{4 \text{ tickets}}$$

A. Choose *two* correct proportions that could be used to solve each problem below. (Be sure there are corresponding units on the top and corresponding units on the bottom.) Do not solve.

1. José earned $45 working at his job for 5 hours. How much would he earn for 7 hours of work?

   (1) $\frac{45}{5} = \frac{7}{h}$ (3) $\frac{5}{45} = \frac{h}{7}$

   (2) $\frac{45}{5} = \frac{h}{7}$ (4) $\frac{5}{45} = \frac{7}{h}$

2. At the Farmers' Market, 9 eggs cost $1.22. How much would 12 eggs cost?

   (1) $\frac{1.22}{9} = \frac{12}{d}$ (3) $\frac{9}{1.22} = \frac{12}{d}$

   (2) $\frac{9}{1.22} = \frac{d}{12}$ (4) $\frac{1.22}{9} = \frac{d}{12}$

3. Jennifer drove for 120 miles on 4.5 gallons of gasoline. How much gasoline would she need to drive 155 miles?

   (1) $\frac{120}{4.5} = \frac{155}{g}$ (3) $\frac{4.5}{120} = \frac{155}{g}$

   (2) $\frac{120}{4.5} = \frac{g}{155}$ (4) $\frac{4.5}{120} = \frac{g}{155}$

4. Samly got 2 base hits for every 5 times he batted. If Samly batted at the same rate, how many hits would he get in 32 at bats?

   (1) $\frac{2}{5} = \frac{b}{32}$ (3) $\frac{5}{2} = \frac{32}{b}$

   (2) $\frac{5}{2} = \frac{b}{32}$ (4) $\frac{2}{5} = \frac{32}{b}$

B. Write a proportion and solve for the unknown in each problem. You may use a calculator.

5. To thicken his sauce, Luis had to add cornstarch. The recipe said to dissolve $1\frac{1}{2}$ teaspoons of cornstarch in 1 tablespoon of water. How much water is needed to dissolve 6 teaspoons of cornstarch?

6. To tile her bathroom, Erika picked a pattern that used 8 blue tiles for every 24 total tiles. If the floor requires 222 tiles, how many blue tiles will Erika need?

C. Use proportions and the chart to solve these problems. You may use a calculator.

7. How many rushing yards would Ohio State have if they rushed at their same rate but played as many games as Miami?

8. How many passing yards would Notre Dame have if their passing rate equaled Miami's?

9. **Write** Do ratios and proportions *determine* the number of yards that a team will gain? Can they be used to *predict* it? What other factors are involved?

Team	Games Played	Rushing Yards	Passing Yards
Ohio State	4	220	424
UCLA	3	210	750
Notre Dame	4	300	720
Miami	5	400	1,500

Understanding Percents

A **percent** represents the number of parts out of 100 parts.

In the box at the right, 40 of the 100 squares are shaded. Therefore, 40% of the squares are shaded and 60% are unshaded.

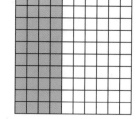

If all of the squares were shaded, then 100% would be shaded.

▶ 100% represents the *whole amount.* (100% = 1)

Percents can be added or subtracted as long as they represent parts of the same whole.

Example 1: If 40 of the 100 squares are shaded, then what percent of the squares are unshaded?

You can figure the percent of unshaded squares by subtracting the percent of shaded squares (40%) from 100%.

100% − 40% = **60%**

Example 2: If 40 of the 100 squares are shaded, and you shade 10 more, what percent of the squares will be shaded?

If 40% of the squares are shaded and you shade 10 more, then you have shaded another 10% of the squares. Add the percents together to find the total percent shaded.

40% + 10% = **50%**

Since 100% stands for the whole amount, a percent less than 100% represents a part of the whole amount. A percent more than 100% represents more than the whole.

Example 3: How much does 125% represent?

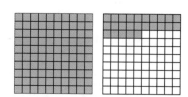

You can rewrite 125% as 100% + 25%. Since 100% is the same as the whole amount, 125% represents the **whole amount plus 25% more.**

A. Find the missing percents in each problem. Use a total of 100%.

1. 30% shaded   67% broken   4% fat   50% in favor of

? % unshaded   ? % unbroken   ? % not fat   ? % against

2. 89% women   30% children   16% alcohol   20% discount

? % men   ? % adults   ? % not alcohol   ? % not discounted

B. Use the diagram to solve problems 3–6.

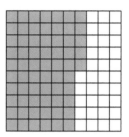

3. If 65 of the 100 squares are shaded, what percent of the squares are shaded?

4. What percent of the squares are unshaded?

5. If you shade 10 more squares, what percent of the squares would be shaded?

6. If you erased the shading from 20 of the 65 shaded squares, what percent would be shaded?

C. Use the chart to solve problems 7–10. Some will take more than one step to solve.

7. What percent of the coins are dimes?

8. What percent of the coins are *not* quarters?

9. What percent of the coins are either pennies or half-dollars?

10. What percent of the coins are *not* quarters or nickels?

Coin Type	Amount
pennies	28
nickels	31
dimes	16
quarters	22
half-dollars	3
Total	100

11. List Make a list of at least five situations in which you have seen percents used.

Decimals, Fractions, and Percents

Decimals, fractions, and percents are all ways to express numbers as amounts other than whole numbers. Decimals use digits to the right of the decimal point to write tenths, hundredths, thousandths, and so on. In fractions, the numerator expresses the number of parts and the denominator expresses the whole. A percent indicates the number of parts out of 100.

Writing a Percent as a Decimal

Example: Write 80% as a decimal.

Step 1
Drop the percent sign and move the decimal point two places *to the left*.

80% = .80
(*Note:* 80% = 80.%)

Step 2
Drop any unnecessary zeros.

.80 = **.8**

> **Tip**
> If the percent is less than 10%, add a 0 in front of the percent before moving the decimal point.
> 8% = 08% = .08

Writing a Decimal as a Percent

Example: What is .3 expressed as a percent?

Step 1
Move the decimal point two places *to the right*, adding any necessary zeros.

.30

Step 2
Add a percent sign.

.30 = **30%**

A. Change the percents below into decimals or whole numbers.

1. 16%   300%   50%   97.8%

2. 3%   8.2%   .7%   1,259%

B. Change the following decimals to percents.

3. .67   .01   .40   3.4

4. 5.06   4.19   .082   23.8

Writing a Percent as a Fraction

Example: Emily finished 25% of her paperwork. What fraction of her paperwork did she finish?

Step 1
Drop the percent sign and write the number with a denominator of 100.

$$25\% = \frac{25}{100}$$

Emily finished $\frac{1}{4}$ of her paperwork.

Step 2
Simplify the fraction if necessary.

$$\frac{25}{100} \div \frac{25}{25} = \frac{1}{4}$$

Writing a Fraction as a Percent

Example: $\frac{4}{5}$ of the registered voters of Barden voted in the mayoral election. What percent of the registered voters voted?

Step 1
Divide the denominator into the numerator.

$$5\overline{)4.0} = .8$$

Step 2
Change the decimal answer into a percent by moving the decimal point two places *to the right*.

.80

Step 3
Add a percent sign.

.80 = 80%

Using a calculator: 4 ÷ 5 =

80% of the registered voters of Barden voted in the election.

 C. Change the following percents into fractions and fractions into percents. You may use a calculator. Round percents to the nearest tenth if needed.

5. 75%   10%   50%   12%

6. $\frac{3}{4}$ $\frac{3}{8}$ $\frac{3}{5}$ $\frac{2}{3}$

 D. Solve the following problems. Change fractions to percents and percents to fractions as requested. You may use a calculator.

7. The Tanakas planted corn on $\frac{5}{8}$ of the acreage on their farm. What percent of the land was planted with corn?

8. $\frac{1}{5}$ of the class received an *A* on the spelling test. What percent of the class received an *A* on the test?

9. Xu Ping scored on 60% of her shots in the recent basketball game. What fraction of her shots were successful?

10. **Chart** Create a chart that lists the percent and decimal equivalents for each of the following fractions that are smaller than 1: halves, thirds, quarters, fifths, and tenths. Use only simplified fractions. Round decimals to the nearest hundredth where needed.

The Percent Equation

A **percent statement** has three parts: the percent, the whole, and the part. In a percent statement you express the part as a percent of the whole.

Percent statement: percent of whole equals part

30% of 120 is 36.

A percent statement can easily be changed into a **percent equation** by replacing the word *of* with a × sign and the word *is* with an = sign. If you use a variable in a percent equation, you can answer one of the following:

- The part is what percent of the whole?
- How much is the whole?
- How much is the part?

Changing a Percent Statement to an Equation

Write a percent equation for each of these examples:

Example 1: 80% of 90 is _____.

Step 1	Step 2	Step 3
Represent the unknown part with a variable.	Change the percent, if there is one, to a decimal.	Replace the word *of* with a × sign and the word *is* with an equal sign.
	80% = .80	.80 of 90 is p.
80% of 90 is p.	.80 of 90 is p.	**$.80 \times 90 = p$**

Example 2: Caitlin made a 20% down payment on a dress. She paid $16. How much was the dress?

Percent statement:
<u>20%</u> of _____ is <u>$16</u>.

Step 1	Step 2	Step 3
Represent the unknown whole with a variable.	Change the percent, if there is one, to a decimal.	Replace the word *of* with a × sign and the word *is* with an equal sign.
20% of w is $16.	.2 of w is $16.	$.2 \times w = \$16$

Example 3: Ramón completed 16 out of 43 forms for the proposed bid. What percent of the forms has Ramón completed?

Percent statement:
_____ of <u>43</u> is <u>16</u>.

Step 1	Step 2	Step 3
Represent the unknown percent with a variable.	Change the percent, if there is one, to a decimal.	Replace the word *of* with a × sign and the word *is* with an equal sign.
n% of 43 is 16.	n% of 43 is 16.	$n\% \times 43 = 16$

A. Change the following percent statements into equations. Use *p* to represent the part, *w* to represent the whole, and *n%* to represent the percent.

1. 45% of 62 is ___.   23% of 134 is ___.   89% of 1,530 is ___.   9% of 431 is ____.

2. 75% of ___ is 65.   6% of ___ is 128.   100% of ___ is 40.   18% of ___ is 850.

3. ___ of 35 is 20.   ___ of 174 is 9.   ___ of 521 is 502.   ___ of 67 is 13.

B. For each problem below, write a percent statement. Then write a percent equation.

4. Maria spends 38% of her salary on rent. If she earns $1,500 per month, how much is her rent?

_____ of _____ is _____.

Equation: _____

5. Ken stopped for gas after driving 65% of the way to his grandparents' house. If he had driven 262 miles, how far is it to his grandparents' house?

_____ of _____ is _____.

Equation: _____

6. In a recent survey, 34 people said they would vote for Sánchez. If 87 people were questioned, what percent intend to vote for Sánchez?

_____ of _____ is _____.

Equation: _____

7. Damon scored 29% of his team's runs. If the team scored 734 runs, how many runs did Damon score?

_____ of _____ is _____.

Equation: _____

For another look
at decimals and percents, turn to page 268.

Making Connections: Circle Graphs

Circle graphs show a whole, or 100% (circle), and its parts (each wedge).

Example: A recent survey questioned residents of Parkville on whether or not to raise money to restore the local art museum. The results of the survey are given in the circle graph to the right.

1. What percent of the residents had no opinion?

2. Write a percent statement and equation for the number of people who said yes.

Museum Restoration Survey
481 people questioned

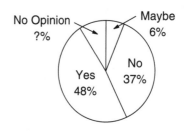

Solving Percent Equations

Solving a percent equation is like solving other equations: your goal is to get the variable alone on one side of the equal sign. In a percent equation, this will always involve multiplying and dividing.

Find the Part in a Percent Equation

Example: Yasmin bought a blouse for $34. If the state sales tax is 5%, how much tax does Yasmin have to pay?

▶ To find the part, multiply the percent by the whole:

$$P = \% \times W$$

Step 1
Write a percent statement.

5% of $34 equals P.

Step 2
Change to a percent equation.

$.05 \times \$34 = P$

Step 3
Multiply the percent by the whole.

$1.7 = \$1.70 = P$

Yasmin has to pay **$1.70** in sales tax on the blouse.

Find the Whole in a Percent Equation

Example: Greg made a 15% down payment on a new car. The amount of the down payment was $1,800. What is the price of the car?

▶ To find the whole, divide the part by the percent:

$$W = \frac{P}{\%}$$

Step 1
Write a percent statement.

15% of W is $1,800.

Step 2
Change to a percent equation.

$.15 \times W = \$1,800$

Step 3
Divide both sides by the percent.

$$\frac{.15 \times W}{.15} = \frac{\$1,800}{.15}$$
$$W = \$12,000$$

The price of the car is **$12,000.**

Find the Percent in a Percent Equation

Example: Of 140 bicycles made at the Peterson plant, 7 were found to be defective. What percent of the bicycles were defective?

▶ To find the percent, divide the part by the whole:

$$\% = \frac{P}{W}$$

Step 1
Write a percent statement.

n% of 140 equals 7.

Step 2
Change to a percent equation.

$n\% \times 140 = 7$

Step 3
Divide both sides by the whole.

$$\frac{n\% \times 140}{140} = \frac{7}{140} = \frac{1}{20}$$
$$n\% = \frac{1}{20} = .05 = 5\%$$

5% of the bicycles were defective.

A. Write and solve percent equations for each statement below.

1. 80% of ___ is 1,200.   ___ % of 55 is 11.   35% of 60 is ___.   15% of ___ is 6.

2. ___% of 18 is 9.   90% of 80 is ___.   24% of ___ is 1,647.   ___ % of 748 is 561.

3. 62% of 114 is ___.   25% of ___ is 2,498.   ___ % of 180 is 120.   7% of 584 is ___.

For another look at decimals and percents, turn to page 268.

B. Solve the following percent problems. You may use a calculator.

4. Eugenia has 17% of her paycheck deducted in taxes. If the amount of the taxes deducted is $43, what was her paycheck before the deduction? Round to the nearest cent.

5. Velma is responsible for signing the restaurant bill for her company's annual dinner. The food bill came to $839. If she leaves a 15% tip, how much was the tip?

6. Kim Manufacturing Company had 1,165 employees last year. This year it hired 128 more workers. By what percent did the company grow? Round to the nearest percent.

7. 90% of Kennedy Middle School students were present on Friday. If 1,350 students were present, what is the total number of students in the school?

C. Use the circle graph to solve problems 8–10. You may use a calculator.

8. How much does the Rosales family spend on rent each month?

9. If the Rosales family's rent changes to $627, what percent of the monthly budget will the rent be?

10. If the family's food expense increases to 25% of take-home pay per month, how much *more* will they spend on food than they do now?

11. **Draw** Create a circle graph of your own monthly expenses. Think of four to six categories for your expenses. One category should be "other" for minor expenses. Try to estimate your expenses to the nearest half, quarter, or sixth.

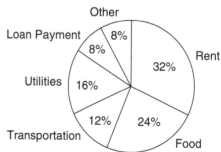

Rosales Family Monthly Budget based on $1,900 take–home pay

- Other 8%
- Loan Payment 8%
- Rent 32%
- Utilities 16%
- Food 24%
- Transportation 12%

Discounts

A discount is a reduction in the original price of an item or service. The reduction is often advertised as a discount percent.

To find the amount of the reduction, multiply the original price by the discount percent. To find the discount price, subtract the discount amount from the original price.

Finding a Discount Price

Example: Steven notices the advertisement at the right in the Sunday newspaper. He decides to take advantage of the sale and buys a sweater originally priced at $34.95. How much did Steven pay for the sweater?

All Sweaters
20% off
Sunday Only!

Step 1
Set up an equation to find the *part,* or the *discount* on the item.

20% of $34.95 is __d__.
.20 × $34.95 = d

Step 2
Solve the equation to determine the amount of the discount.

.20 × $34.95 = d
$6.99 = d

Calculator: 34.95 × 20% = 6.99

Step 3
Subtract the discount from the original price.

$34.95 − $6.99 = $27.96
original price   discount price

Calculator: 34.95 − 6.99 = 27.96

The sweater now costs **$27.96,** a savings of $6.99 off the original price.

A. Find the savings and discount price in each problem. Round to the nearest cent, if necessary. You may use a calculator.

1. A pair of shoes with an original price of $37.45 at an 8% discount.

   Savings: _____

   Discount price: _____

2. A $388 mattress discounted at 25% off.

   Savings: _____

   Discount price: _____

3. A picture frame with a regular price of $12.98 at a 15% discount.

   Savings: _____

   Discount price: _____

4. A discontinued model of a television set at 15% off the regular price of $562.15.

   Savings: _____

   Discount price: _____

Another Way to Look at Discounts

To figure a discount price, first you find the savings. Then you subtract the savings from the regular price. It is possible to find discounted prices in a slightly easier way—if you remember one important fact:

▶ 100% − % of Discount = % Paid

Example: Steven is buying a sweater at 20% off the regular price of $34.95. How much does the sweater cost?

Step 1
Subtract the discount percent from 100%.

Step 2
Set up the equation to multiply the percent by the original price.

Step 3
The answer is the discounted price.

100% − 20% = 80%

.80 × $34.95 = d

$27.96 = d
Calculator: 34.95 × 80% = 27.96

The sweater now costs **$27.96,** the same result calculated in the example on page 106.

This second method is simpler because you can do the first step in your head (subtracting the discount percent from 100%). You can find the savings by taking the regular price and subtracting the discounted price.

B. Find the discounted price using the second method. Round to the nearest cent, if necessary. You may use a calculator.

5. A stereo with an original price of $399 with a 12% discount.

   Discount price: _____

6. An atlas selling for $46.31 with a 30% discount.

   Discount price: _____

Making Connections: Using a Calculator to Find Discounts

Calculators make the process of finding a discounted price very simple. Just enter the regular price, press the minus sign ⊟, the discount percent and the ⟨%⟩. The display shows the discount price. (On some calculators, you also have to press ⟨=⟩.)

Use the process described here to solve problems 5 and 6 in this lesson.

Two-Step Percent Problems

In this unit, you've just practiced finding the part, the percent, and the whole. In many math problems and real-life situations, you must solve a percent equation, then use the information to find another answer.

Solving Two-Step Percent Problems

Example 1: Mario makes $9.00 per hour. After one year on the job, he gets a 6% raise. What is Mario's hourly wage after the raise?

Step 1	**Step 2**	**Step 3**	**Step 4**
Set up the percent equation.	Solve for the unknown part.	Decide what operation is needed for the second step.	Solve.
6% of $9.00 is ___.	$.06 \times 9.00 = p$	*Add the increase in hourly pay to the old hourly pay to find the new hourly pay.*	$0.54 + $9.00 = $9.54
$.06 \times 9.00 = p$	$.54 = p$		

Mario now earns **$9.54** per hour.

> ### Tip
> An easier way to increase an amount by a percent is to add 100% to the percent increase, change to a decimal, and multiply. (You used a similar technique in the previous lesson when you subtracted the percent from 100% to find the discount price.)
>
> | | |
> |---|---|
> | Add 100%. | 6% + 100% = 106% |
> | Change to a decimal. | 106% = 1.06 |
> | Multiply. | $9.00 \times 1.06 =$ **$9.54** |

Example 2: Diane pays $60 a month, or 2% of her paycheck, in interest on a loan. How much does Diane earn *in a year*?

Step 1	**Step 2**	**Step 3**	**Step 4**
Set up the percent equation.	Solve for the unknown whole.	Decide what operation is needed for the second step.	Solve.
2% of ___ is $60.	$.02 \times w = 60$	*Multiply monthly earnings by 12 to find the yearly earnings.*	$3,000 \times 12 = $36,000
$.02 \times w = 60$	$w = 60 \div .02$		
	$w = $3,000		

Diane earns **$36,000** a year.

A. Solve the following two-step percent problems. Be sure you write and solve a percent equation and also add, subtract, multiply, or divide.

1. Jim averaged 10.4 yards per catch in football last season. In the homecoming game he averaged 12% more per catch. If Jim caught 8 passes, how many yards did he gain? (Round to the nearest yard.)

2. Katrina gets a 12% commission on the clothing she sells. Last month she received a commission check for $240. How much clothing did Katrina sell *per day,* if she worked 20 days during the month? (Round your answer to the nearest cent.)

3. José sold 230 cars last year. This year he sold 275 cars. By what percent did José increase his car sales? (Round your answer to the nearest percent.)

4. The Schmidt family made a 10% down payment on a house. The amount of the down payment was $14,000. How much does the family still owe on the house?

B. Use the discount table to solve problems 5 and 6. Use the tax table to solve problems 7 and 8. You may use a calculator. Round your answers to the nearest cent.

Discounts	
$0.00 - $25.00	5%
$25.01 - $50.00	7.5%
$50.01 - $75.00	10%
$75.01 and up	12.5%

Tax Table	
Danville	8.25%
Frankfort	6.25%
Villa Park	8.00%
Yorktown	7.75%

5.

Find the discounted price if the original price is $34.80.

7.

Find the total cost in Frankfort if the discounted price is $125.67.

6.

Find the discounted price if the original price is $18.45.

8.

Find the total cost in Yorktown if the discounted price is $96.38.

9. **Multiple Solutions** Use both tables above to figure the total cost of a dress purchased in Danville with a regular price of $55.85. Figure the tax amount *after* subtracting the discount amount. Use both methods described on pages 106–107.

Unit 3 Review

A. Solve the problems below. Simplify answers to lowest terms and change improper fractions to whole or mixed numbers.

1. $\frac{5}{9} + \frac{1}{9}$ $\qquad$ $\frac{3}{8} + \frac{2}{3}$ $\qquad$ $\frac{7}{10} - \frac{3}{10}$ $\qquad$ $\frac{5}{8} - \frac{1}{6}$

2. $3\frac{1}{4} - 1\frac{1}{3}$ $\qquad$ $\frac{4}{5} \times \frac{2}{3}$ $\qquad$ $\frac{3}{4} \times \frac{2}{3}$ $\qquad$ $4\frac{1}{2} \times \frac{2}{9}$

3. $\frac{1}{6} \div \frac{2}{3}$ $\qquad$ $\frac{5}{8} \div \frac{5}{9}$ $\qquad$ $2\frac{3}{4} \div \frac{1}{8}$ $\qquad$ $\frac{15}{16} \div 3$

B. Choose the best answer to each word problem below.

4. Put the following fractions in order from *greatest* to *least*.

   A: $\frac{1}{4}$ $\quad$ B: $\frac{1}{3}$ $\quad$ C: $\frac{3}{8}$ $\quad$ D: $\frac{1}{2}$ $\quad$ E: $\frac{2}{5}$

   (1) E, D, C, B, A
   (2) C, E, B, D, A
   (3) D, E, C, B, A
   (4) D, C, B, E, A
   (5) D, A, C, B, E

5. Taj spends 6 hours a day at school. Which of the numbers below does *not* express the portion of a day that Taj is at school?

   (1) $\frac{1}{6}$
   (2) $\frac{1}{4}$
   (3) 25%
   (4) .25
   (5) 1:4

6. Choose the ratio that best describes the comparison of circles to squares shown below.

   (1) 3:4
   (2) 4:7
   (3) 3:7
   (4) 4:3
   (5) 7:4

7. Maria puts 5% of her paycheck in a savings account each week. She deposits $5.25 each week. Which of the following percent equations can you use to find out how much Maria earns per week?

   (1) $5.25 \times .05 = w$
   (2) $5.25 \div w = .05$
   (3) $.05 \times 5.25 = w$
   (4) $.05 \times w = 5.25$
   (5) $.05 \times 5.25 = w$

C. Solve the following problems. Round to the nearest cent or percent, if necessary. You may use a calculator.

SALE Save **20%** on socks

Today Only
All of our fine sheets
15% off
the regular price

Going out of Business
30% off
Any CD
or Tape

8. Jesse wants to buy a pair of socks normally priced at $3.75. How much money will he save off the original price if he buys the socks at 20% off?

9. Brian normally buys sheets for $35 at Brown's. The same sheets are on sale at Federal Department Store at 15% off. The regular price at Federal is $40. Which store has the better buy?

10. Abdul bought a CD normally selling for $12.30. The store was supposed to discount the CD at 30% off. When he got home he looked at his cash register receipt. The store had used $8.20 as the discounted price. What percent discount did Abdul get?

11. Susannah buys a CD regularly priced at $14.50 and a tape regularly priced at $6.75. How much will the items cost if she buys them on sale at 30% off?

12. Lana buys 2 pairs of socks at 20% off. At full price, one pair costs $2.86 and the other pair costs $4.75. What will the total discounted price be for the socks?

13. Yoko bought a pair of pillowcases that cost her a total of $24.90. She got the pillowcases on sale at 15% off and paid a state sales tax of 8%. What was the original price of the pillowcases? (*Hint:* First find the price before the tax was added. This was the discount price.)

D. Use the circle graph at the right to answer the following questions. Round your answer to the nearest whole number. You may use a calculator.

14. How many compact cars were sold?

15. What was the total number of vans and trucks sold?

16. How many nonluxury vehicles were sold?

Hollywood Motor Company
(382 vehicles sold)

Other 8%
Luxury 13%
Compacts 38%
Trucks 24%
Vans 17%

Working Together

Work with a partner to create circle graphs that illustrate how you each spend a typical day. Use four to six main categories (including "other"). Estimate the sizes of the graph's sections.

Data and Measurement

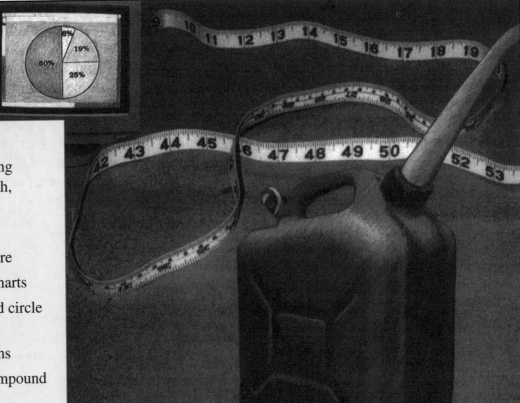

Skills

Adding and subtracting English units of length, weight, and capacity

Using metric units

Measuring temperature

Reading tables and charts

Reading bar, line, and circle graphs

Using scatter diagrams

Using simple and compound probability

Tools

Rulers, cups, and spoons

Scales and meters

Computer spreadsheets

Problem Solvers

Using more than one data source

Seeing trends and making predictions

Applications

Figuring distance, rate, and time

Using mean, median, and mode

The term **data** describes a collection of facts. When the facts are numerical, we can use mathematics to analyze and draw conclusions from the data.

Most often, numerical data is in the form of **measurements.** In this unit you will calculate using basic units of length, weight, and capacity. You will gain more practice using time and money.

This unit also introduces the concepts of simple and compound probability. In mathematics, **probability** deals with how likely events are to occur. In working with probability, you will make predictions based on the laws of chance.

Measurement, data analysis, and probability provide information on which we base many of our practical decisions in life. In this unit you will practice working with data, measurement, and probability.

When Do I Measure and Use Data?

Each of the following situations uses data, measurement, or both. Check the experiences you have had.

- ☐ using a ruler, yardstick, or tape measure
- ☐ measuring quantities in a recipe
- ☐ using an index to find information in a book
- ☐ reading a table to find how much medicine to give a child
- ☐ using a tax table to find out how much you owe in taxes
- ☐ choosing the best strategy for winning a game

If you've done any of those things, you've used data analysis and measurement. Describe some of your experiences below.

1. The last time you used a ruler, yardstick, meterstick, or tape measure, what were you measuring? Describe any calculations you performed with the measurements.

   _____

   _____

2. Think of a game that you are good at in which chance is involved. Describe any strategies you use that help you to win.

   _____

   _____

3. When you cook, do you measure, or do you estimate? Which units of measure are you likely to estimate?

   _____

   _____

4. Sales clerks in small businesses often use a table to look up sales tax. The clerk could also use percent to figure out the tax. Have you ever used a table to look up information on a job? Would you rather use a table or do the math? Under what circumstances do you think tables are a good idea?

   _____

   _____

Talk About It

Blake had the following grades in a math class: 95, 92, 64, 92, and 97. The range for an *A* is 90–100. Because Blake's average was 88, the teacher gave him a *B*. Blake felt the grade was unfair. What do you think? Can you think of any other ways the teacher could have figured the grade? Discuss your ideas with others.

Throughout this unit, you may use a calculator wherever that would be useful.

English Units of Length

The measuring system most often used in the United States is called the **English system.** The most often used English units of length are shown below.

Unit	Sample Use	Comparing Units
inch (in.)	11 in.	$1 \text{ in.} = \frac{1}{12} \text{ ft.}$
foot (ft.)	6 ft.	$1 \text{ ft.} = 12 \text{ in.}$ $1 \text{ ft.} = \frac{1}{3} \text{ yd.}$
yard (yd.)	10 20 30 40 50 40 30 20 10 — 100 yd. —	$1 \text{ yd.} = 3 \text{ ft.}$ $1 \text{ yd.} = 36 \text{ in.}$
mile (mi.)	Next Exit 3 Miles	$1 \text{ mi.} = 1{,}760 \text{ yd.}$ $1 \text{ mi.} = 5{,}280 \text{ ft.}$

Sometimes lengths or distances need to be written in larger units.
To change to a larger unit, *divide*.

$$\frac{\text{smaller}}{\text{units}} \xrightarrow[\text{divide}]{\div} \frac{\text{larger}}{\text{units}}$$

Converting to Larger Units

Example: Marla wonders if the couch she wants to buy will fit against a wall that is 7 ft. long. The couch is 76 in. long. Convert its length to feet and inches.

Step 1
Set up the problem.

Step 2
Estimate first by rounding to the nearest multiple of the smaller unit (12 in.).

Step 3
Divide by the number of smaller units.
(12 in. = 1 ft.)

Step 4
Write the whole number as the larger unit and the remainder as the smaller unit.

76 in. = __ ft. __ in.

76 in. ≈ 72 in.[*]
72 in. ÷ 12 = 6 ft.

76 in. ÷ 12 = 6 R4

76 in. = 6 ft. 4 in.

Check: Your answer of **6 ft. 4 in.** is close to the estimate of 6 ft.
The couch will fit against the wall.

---

[*]*Remember:* The symbol ≈ means "is approximately equal to."

A. Change each length to the larger unit indicated. Estimate first.

1. 28 in. ≈ ___ ft.   47 in. ≈ ___ ft.   16 ft. ≈ ___ yd.

    28 in. = ___ ft. ___ in.   47 in. = ___ ft. ___ in.   16 ft. = ___ yd. ___ ft.

    Hint: 1 foot = ? inches

2. 23 ft. ≈ ___ yd.   81 in. ≈ ___ yd.   110 in. ≈ ___ yd.

    23 ft. = ___ yd. ___ ft.   81 in. = ___ yd. ___ in.   110 in. = ___ yd. ___ in.

To change to a smaller unit, *multiply.*

$$\text{larger units} \xrightarrow[\text{multiply}]{\times} \text{smaller units}$$

Converting to Smaller Units

Example: Roberta needed $6\frac{1}{4}$ feet of material for a window covering. How many inches of material does she need?

Step 1
Set up the problem.

$6\frac{1}{4}$ ft. = __ in.

Step 2
Estimate first by rounding to the nearest whole number.

$6\frac{1}{4}$ ft. ≈ 6 ft.

6 ft. × 12 = 72 in.

Step 3
Multiply by the number of smaller units.

$6\frac{1}{4} \times 12 = \frac{25}{\cancel{4}} \times \frac{\cancel{12}^{\,3}}{1} = 75$ in.

Check: Your answer of **75 in.** is close to the estimate of 72 in.

B. Change each length to the smaller unit indicated. Estimate first.

3. 3 ft. ≈ ___ in. $5\frac{1}{6}$ ft. ≈ ___ in. $1\frac{1}{3}$ yd. ≈ ___ ft.

    3 ft. = ___ in. $5\frac{1}{6}$ ft. = ___ in. $1\frac{1}{3}$ yd. = ___ ft.

4. $4\frac{3}{4}$ yd. ≈ ___ in. $1\frac{1}{5}$ mi. ≈ ___ yd. $2\frac{5}{6}$ mi. ≈ ___ ft.

    $4\frac{3}{4}$ yd. = ___ in. $1\frac{1}{5}$ mi. = ___ yd. $2\frac{5}{6}$ mi. = ___ ft.

C. Solve the following word problems. Estimate first.

5. George needed 129 feet of fencing to protect his pool. The hardware store sells the fencing by the yard. How many yards of fencing is 129 feet?

6. Lila must mail several boxes of personnel forms to the branch office. She estimates she needs 233 inches of packing tape. If the store sells tape by the whole yard only, how many yards of tape does Lila need?

7. Marcus is the tallest person in his school at $6\frac{1}{2}$ feet tall. How many inches tall is Marcus?

8. Multiple Answers Rashaad passed for 12,534 yards during his football career. List his passing yards in as many different units of length as you can (in., ft., mi.).

Working with Length

When working with lengths, add or subtract each unit separately.
Estimate first. To estimate, round to the nearest whole number of
larger units, and then add or subtract.

Adding Units of Length

Example: The manager of Go-Mart wants a display measuring 4 ft. 5 in. wide placed next
to a display measuring 2 ft. 8 in. wide. How much space will the two displays take up?

Step 1	**Step 2**	**Step 3**	**Step 4**
Set up the problem.	Estimate first by rounding to the nearest whole number of larger units.	Add the units separately.	Regroup the smaller unit and combine with the larger unit, if necessary.

Step 1:
$$4 \text{ ft. } 5 \text{ in.}$$
$$+ \ 2 \text{ ft. } 8 \text{ in.}$$

Step 2:
4 ft. 5 in. ≈ 4 ft.
2 ft. 8 in. ≈ 3 ft.
4 ft. + 3 ft. = 7 ft.

Step 3:
$$4 \text{ ft. } \ 5 \text{ in.}$$
$$+ \ 2 \text{ ft. } \ 8 \text{ in.}$$
$$6 \text{ ft. } 13 \text{ in.}$$

Step 4:
6 ft. + 13 in. =
6 ft. + 1 ft. + 1 in. =
7 ft. 1 in.

Check: Your answer of **7 ft. 1 in.** is close to the estimate of 7 ft.

For another look at units of measurement, turn to page 269.

Subtracting Units of Length

Example: The owners of Go-Mart want to install a 12-ft.-long cooler. The space they
have in mind is 10 ft. 8 in. long. How much more space is needed for the cooler to fit?

Step 1	**Step 2**	**Step 3**	**Step 4**
Set up the problem.	Estimate first by rounding to the nearest whole number of larger units.	Regroup from the larger unit, if necessary.	Subtract each unit separately.

Step 1:
$$12 \text{ ft.}$$
$$- \ 10 \text{ ft. } 8 \text{ in.}$$

Step 2:
12 ft. − 11 ft. = 1 ft.

Step 3:
12 ft. = 11 ft. 12 in.

Step 4:
$$11 \text{ ft. } 12 \text{ in.}$$
$$- \ 10 \text{ ft. } \ 8 \text{ in.}$$
$$1 \text{ ft. } \ 4 \text{ in.}$$

Check: Your answer of **1 ft. 4 in.** is close to the estimate of 1 ft.

A. Add or subtract the following lengths. Estimate first. Regroup when necessary.

1.
6 ft. 4 in.	2 ft. 10 in.	7 ft. 10 in.	4 yd. 2 ft.
+ 3 ft. 5 in.	+ 5 ft. 6 in.	+ 6 ft. 11 in.	+ 2 yd. 2 ft.

2.
4 ft. 7 in.	5 ft. 6 in.	9 ft. 1 in.	6 yd. 1 ft.
− 1 ft. 3 in.	− 2 ft. 8 in.	− 4 ft. 10 in.	− 1 yd. 2 ft.

Multiplying Units of Length

Example: Anna needs a wooden dowel that she can cut into 5 pieces. Each piece must be 1 ft. 3 in. How long should the wooden dowel be?

Step 1
Set up the problem.

1 ft. 3 in.
× 5
———

Step 2
Estimate by rounding to the nearest whole number of larger units.

1 ft. 3 in. ≈ 1 ft.
1 ft. × 5 = 5 ft.

Step 3
Multiply the number of each unit separately.

1 ft. 3 in.
× 5
———
5 ft. 15 in.

Step 4
Simplify, if possible.

5 ft. 15 in. =
5 ft. + 1 ft. + 3 in. =
6 ft. 3 in.

Check: Your answer of **6 ft. 3 in.** is reasonably close to the estimate of 5 ft.

Dividing Units of Length

Example: If you divide a 13-ft.-3-in. board into 3 equal pieces, how long is each piece?

Step 1
Set up the problem.

$3\overline{)13\text{ ft. 3 in.}}$

Step 2
Estimate by rounding to the nearest whole number of larger units.

13 ft. 3 in. ≈ 13 ft.
13 ft. ÷ 3 = $4\frac{1}{3}$ ft.

Step 3
Divide the larger unit.

$$\begin{array}{r} 4 \\ 3\overline{)13\text{ ft.}} \\ -12 \\ \hline 1 \end{array}$$

Remainder
1 ft. = 12 in.

Step 4
Regroup and add any remainder to the smaller unit. Divide the smaller unit.

12 in. + 3 in. = 15 in.

$$\begin{array}{r} 5 \\ 3\overline{)15\text{ in.}} \\ -15 \\ \hline 0 \end{array}$$

Step 5
Combine the amounts from steps 3 and 4 to get the answer.

4 ft. + 5 in. =
4 ft. 5 in.

Check: Your answer of **4 ft. 5 in.** is close to the estimate of $4\frac{1}{3}$ ft.

B. Multiply or divide the following lengths. Estimate first.

3.
2 ft. 4 in.
× 3
———

5 ft. 2 in.
× 5
———

4 ft. 9 in.
× 4
———

7 ft. 10 in.
× 6
———

4. $2\overline{)4\text{ ft. 8 in.}}$ $4\overline{)5\text{ ft. 4 in.}}$ $5\overline{)12\text{ ft. 1 in.}}$ $4\overline{)6\text{ yd. 2 ft.}}$

C. Solve the following word problems. Estimate first.

5. A.J. plants 4 bushes in a 6-ft.-9-in.-long flower bed. Two of the bushes are at opposite ends. If the bushes are equally spaced, how far is each from the next?

6. A junior basketball team has 3 guards. Their heights are 4 ft. 2 in., 5 ft. 4 in., and 5 ft. 9 in. What is the average height of the guards? (*Hint:* Add and divide.)

7. Tanya bought 3 computer cables. Each cable was 6 yd. 1 ft. What was the total length of the cables?

8. **Discover** Find 3 whole numbers that divide into 3 ft. 8 in. and give an answer that does not include fractions of an inch. Then find 3 whole numbers by which to multiply 3 ft. 8 in. to get a whole number of feet.

Measuring Capacity

Units of **capacity** measure the space occupied by substances that can be poured (liquid or granular materials, for example). The most often used English units of capacity are shown in the table.

Unit	Sample Use	Comparing Units
fluid ounce (fl. oz.)	1 fl. oz.	$1 \text{ fl. oz.} = \frac{1}{8} \text{ c.}$
cup (c.)	1 cup 8 fl. oz.	1 c. = 8 fl. oz.
pint (pt.)	Milk	1 pt. = 2 c.
quart (qt.)	motor oil	1 qt. = 2 pt.
gallon (gal.)	GASOLINE	1 gal. = 4 qt.

Converting to Larger Units

Example: Jim has 47 fl. oz. of juice left in a carton. Express this amount in cups and fluid ounces.

Step 1
Set up the problem.

Step 2
Estimate first by rounding to the nearest multiple of the smaller unit (8 fl. oz.).

Step 3
Divide by the number of smaller units. (8 fl. oz. = 1 c.)

Step 4
Write the whole number as the larger unit and the remainder as the smaller unit.

47 fl. oz. = __ c. __ fl. oz.

47 fl. oz. ≈ 48 fl. oz.
48 fl. oz. ÷ 8 = 6 c.

47 fl. oz. ÷ 8 =
5 R7

47 fl. oz. = 5 c. 7 fl. oz.

Check: Your answer of **5 c. 7 fl. oz.** is close to the estimate of 6 c.

Converting to Smaller Units

Example: Soo Jung has $1\frac{1}{4}$ qt. of applesauce. She needs 38 fl. oz. Does she have enough?

Step 1
Set up the problem.

Step 2
Find out how many fluid ounces there are in 1 qt.

Step 3
Estimate first.

Step 4
Multiply by the number of smaller units.

$1\frac{1}{4}$ qt. = __ fl. oz.

1 qt. = 2 pt.
2 pt. = 4 c.
4 c. = 32 fl. oz.

$1\frac{1}{4}$ qt. ≈ 1 qt.
1 qt. = 32 fl. oz.

$1\frac{1}{4} \times 32 = \frac{5}{4} \times \frac{32}{1} =$ 40 fl. oz.

Yes, Soo Jung has enough applesauce (40 fl. oz.).

Check: 40 fl. oz. is close to the estimate of 32 fl. oz.

A. Change units as indicated. Estimate first.

1. 24 fl. oz. ≈ ___ c.   5 c. ≈ ___ pt.   7 pt. ≈ ___ qt.   14 qt. ≈ ___ gal.

   24 fl. oz. = ___ c.   5 c. = ___ pt. ___ c.   7 pt. = ___ qt. ___ pt.   14 qt. = ___ gal. ___ qt.

2. $2\frac{1}{8}$ c. ≈ ___ fl. oz.   10 pt. ≈ ___ c. $3\frac{3}{4}$ gal. ≈ ___ qt. $5\frac{1}{2}$ qt. ≈ ___ c.

   $2\frac{1}{8}$ c. = ___ fl. oz.   10 pt. = ___ c. $3\frac{3}{4}$ gal. = ___ qt. $5\frac{1}{2}$ qt. = ___ c.

Adding or Subtracting Units of Capacity

Example: Eileen has a 12-gal.-2-qt. container. To fill the container, she adds 8 gal. 3 qt. of water. How much water was in the container before she filled it?

Step 1
Set up the problem.

12 gal. 2 qt.
− 8 gal. 3 qt.

Step 2
Estimate first.

12 gal. − 8 gal. = 4 gal.

Step 3
Regroup, if necessary.

12 gal. 2 qt. =
11 gal. + 4 qt. + 2 qt. =
11 gal. 6 qt.

Step 4
Add or subtract the units separately. Simplify, if necessary.

11 gal. 6 qt.
− 8 gal. 3 qt.
3 gal. 3 qt.

Check: Your answer of **3 gal. 3 qt.** is close to the estimate of 4 gal.

B. Add or subtract the following capacities. Estimate first.

3.
   2 c. 6 fl. oz.
   + 3 c. 3 fl. oz.

   2 pt. 1 c.
   + 2 pt. 1 c.

   1 qt. 1 pt.
   + 3 qt. 1 pt.

   4 gal. 3 qt.
   + 3 gal. 2 qt.

4.
   6 c. 5 fl. oz.
   − 3 c. 2 fl. oz.

   8 c. 4 fl. oz.
   − 4 c. 7 fl. oz.

   5 pt. 0 c.
   − 2 pt. 1 c.

   10 gal. 1 qt.
   − 3 gal. 3 qt.

C. Solve the following word problems. Estimate first.

5. Yi had a 3-qt.-1-pt. bottle of water. He used 2 qt. 1 pt. How much water was left?

6. To make instant cider, mix 6–8 fl. oz. of hot water with a pouch of mix. If Dave adds 3 c. of water to 4 pouches of mix, is that enough water?

7. Amanda used 3 c. of milk from a 1-gal. container. How much milk is left?

8. **Chart** Create a chart listing the number of fluid ounces and cups in a pint, quart, and gallon.

Using Rulers, Cups, and Spoons

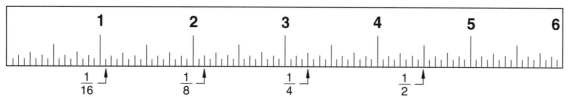

A common measuring tool found around many homes is the English ruler. A 6-in. ruler is pictured below.

Each inch is divided into $\frac{1}{16}$-, $\frac{1}{8}$-, $\frac{1}{4}$-, and $\frac{1}{2}$-in. fractions. The $\frac{1}{16}$-in. marks are shortest, the $\frac{1}{8}$-in. marks are longer, and so on.

Reading Rulers

Example: Maria is building a chair in her wood-shop class. She needs to measure the distance from the edge of the wood to the hole as shown below.

Step 1
Line up the edge of the wood with the left edge of the ruler.

Step 2
Find the whole number of inches at or to the left of the hole.

3 in.

Step 3
Find the fraction mark closest to the edge of the hole.

$\frac{1}{8}$-in. mark

Step 4
Count the number of $\frac{1}{8}$-inch marks starting at the last inch mark.

Count $\frac{1}{8}, \frac{2}{8}, \frac{3}{8}$.

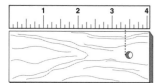

The distance from the edge to the hole is **$3\frac{3}{8}$ in.**

Note: The $3\frac{3}{8}$-in. length shown here is also equal to $3\frac{6}{16}$ in. Unless a specific fractional amount is asked for, give amounts in simplest terms.

A. What are the lengths marked by the letters below?

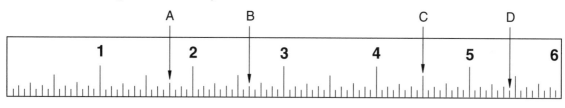

1. A = _____   B = _____   C = _____   D = _____

B. Based on part A, determine each of the following distances.

2. the distance from A to B

the distance from B to C

the distance from B to D

the distance from C to D

Cups and Spoons

Pictured below are an English measuring cup, tablespoon, and teaspoon. Each of these tools is used to measure standard units in cooking.

1 c. = 8 fl. oz.

($\frac{1}{2}$ c. = 4 fl. oz.)

1 tablespoon (tbsp.) = $\frac{1}{2}$ fl. oz.

1 tbsp. = 3 teaspoons

1 teaspoon (tsp.) = $\frac{1}{6}$ fl. oz.

C. **How much liquid is in the cups pictured below?**

3. _____ c. (fraction)

_____ fl. oz.

4. _____ c. (fraction)

_____ fl. oz.

D. **Solve each measurement problem below.**

5. Melvin can't find his measuring cup. He needs to add $\frac{1}{4}$ c. of water to his pastry recipe. How many tablespoons of water should Melvin measure instead?

6. **Explain** Corrine has 3 fl. oz. of milk in a small container. If she uses 5 tbsp. making hot cereal and 3 tsp. in her coffee, will she have any milk left over? If so, how much? Explain how you found your answer.

Measuring Weight

Weight is a measure of how heavy an object is. In the English system, weight is measured in the three basic units shown in the table.

Unit	Sample Use	Comparing Units
ounce (oz.)		$1 \text{ oz.} = \frac{1}{16} \text{ lb.}$
pound (lb.)		$1 \text{ lb.} = 16 \text{ oz.}$
ton (t.)		$1 \text{ t.} = 2{,}000 \text{ lb.}$

Converting to Larger Units

Example: Ibrahim bought a bag of candy weighing 45 oz. What is that weight in pounds and ounces?

Step 1
Set up the problem.

Step 2
Estimate first by rounding to the nearest multiple of the smaller unit (16 oz.).

Step 3
Divide by the number of smaller units.
(16 oz. = 1 lb.)

Step 4
Write the whole number as the larger unit and the remainder as the smaller unit.

45 oz. =
__ lb. __ oz.

45 oz. ≈ 48 oz.
48 oz. ÷ 16 = 3 lb.

45 oz. ÷ 16 =
2 R13

45 oz. = 2 lb. 13 oz.

Check: Your answer of **2 lb. 13 oz.** is close to the estimate of 3 lb.

Converting to Smaller Units

Example: Heidi needed 60 oz. of beef to serve her guests at a dinner party. She bought $3\frac{3}{4}$ lb. of meat. Was that enough?

Step 1
Set up the problem.

Step 2
Estimate first (1 lb. = 16 oz.).

Step 3
Multiply by the number of smaller units.

$3\frac{3}{4}$ lb. = __ oz.

$3\frac{3}{4}$ lb. ≈ 4 lb.
4 lb. × 16 = 64 oz.

$3\frac{3}{4} \times 16 = \frac{15}{\overset{}{\underset{1}{4}}} \times \frac{\overset{4}{16}}{1} = 60$ oz.

Yes, Heidi has just enough meat (60 oz.).

Check: 60 oz. is close to the estimate of 64 oz.

A. Change each amount into larger or smaller units as indicated. Estimate first.

1. 32 oz. ≈ __ lb.   40 oz. ≈ __ lb.   92 oz. ≈ __ lb.   4,500 lb. ≈ __ t.

   32 oz. = __ lb.   40 oz. = __ lb. __ oz.   92 oz. = __ lb. __ oz.   4,500 lb. = __ t. __ lb.

2. $1\frac{1}{2}$ lb. ≈ ___ oz. $2\frac{1}{4}$ lb. ≈ ___ oz.   3 t. ≈ ___ lb. $4\frac{3}{4}$ t. ≈ ___ lb.

   $1\frac{1}{2}$ lb. = ___ oz. $2\frac{1}{4}$ lb. = ___ oz.   3 t. = ___ lb. $4\frac{3}{4}$ t. = ___ lb.

Adding or Subtracting Units of Weight

Example: Cobi roasted 2 chickens. The first weighed 4 lb. 10 oz. The second weighed
3 lb. 14 oz. What was the chickens' combined weight?

Step 1	**Step 2**	**Step 3**	**Step 4**
Set up the problem.	Estimate first.	Add or subtract the units separately.	Regroup or simplify the answer, if necessary.

Step 1:
    4 lb. 10 oz.
  + 3 lb. 14 oz.

Step 2:
  5 lb. + 4 lb. = 9 lb.

Step 3:
    4 lb. 10 oz.
  + 3 lb. 14 oz.
    7 lb. 24 oz.

Step 4:
  7 lb. 24 oz. =
  7 lb. + 1 lb. 8 oz. =
  8 lb. 8 oz.

Check: Your answer of **8 lb. 8 oz.** is close to the estimate of 9 lb.

B. Add or subtract the following weights. Estimate first.

3.
     2 lb. 3 oz.
  + 1 lb. 8 oz.

     6 lb. 14 oz.
  + 4 lb. 11 oz.

     2 t. 152 lb.
  + 1 t. 576 lb.

     3 t. 1,500 lb.
  + 5 t. 1,250 lb.

4.
     5 lb. 12 oz.
  − 3 lb. 8 oz.

     7 lb. 3 oz.
  − 5 lb. 12 oz.

     12 lb. 14 oz.
  − 6 lb. 15 oz.

     4 t.   800 lb.
  − 1 t. 1,200 lb.

C. Solve the following word problems. Estimate first.

5. Trevor used about half of a 5-lb. bag of potatoes (2 lb. 6 oz.). How much did the remaining potatoes weigh?

6. Tyus can carry $1\frac{1}{2}$ t. of cargo in his truck. He loads the truck with 1,756 lb. of bricks. How many more pounds of bricks can Tyus put in his truck?

7. McGarity's Bar and Grill had 35 lb. 3 oz. of steak left over from last night. They received a delivery of 14 lb. 15 oz. today. How much steak do they have altogether?

8. Keiko is using a forklift to load shipping crates on a freight elevator. Each crate weighs 220 lb. The elevator can hold 1 t. safely. Can she safely load 9 crates?

9. **Discuss** Which English weight units would you use to measure a truck? A man? A box of cereal?

Using Metric Units

The most commonly used measurement system is the metric system. The standard metric unit of length is the **meter.** Some of the most common metric units of length are shown here with similar English units.

Metric Units of Length			
1 millimeter (mm)	▏	1 meter (m)	
$\frac{1}{16}$ inch (in.)	▏	1 yard (yd.)	
	actual size		to scale
1 centimeter (cm)	▃	1 kilometer (km)	
1 inch (in.)	▃▃	1 mile (mi.)	
	actual size		to scale

10 millimeters = 1 centimeter	100 centimeters = 1 meter
1,000 millimeters = 1 meter	1,000 meters = 1 kilometer

A. Choose the units you would use from the English and metric systems to measure each of the following items.

1. _____

_____

2. _____

_____

3. _____

_____

Look at the chart above. Notice that the metric system is based on tens. For example, 10 millimeters equal 1 centimeter and 100 centimeters equal 1 meter.

When you are converting metric measurements, remember that the metric system is based on tens.

▶ To change a larger metric unit to a smaller unit, *multiply by the number of smaller units* in the larger unit.

Example: 1.5 km = _____ m
Think: 1.5 × 1,000 = 1.500 = **1,500 m**
                      Move decimal point 3 places.

B. Change the following measurements to the units indicated.

4. 19.4 cm = _____ mm   **6.** 20 km = _____ m   **8.** .5 km = _____ m

5. 8.7 m = _____ cm   **7.** 2.9 cm = _____ mm   **9.** .64 m = _____ cm

Some common metric units of weight and capacity are shown below. The standard metric unit of weight is the **gram.** The standard metric unit of capacity is the **liter.** Remember, capacity measures liquid or granular amounts.

Metric Units of Weight
1 milligram (mg) about the weight of a grain of salt
1 gram (g) about the weight of a pencil eraser
1 kilogram (kg) a little over 2 pounds

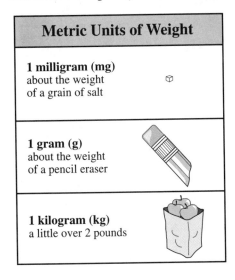

Metric Units of Capacity
1 milliliter (ml) about the size of a sugar cube
1 liter a little more than 1 quart

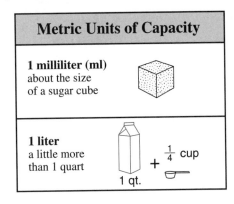

1,000 milligrams = 1 gram
1,000 grams = 1 kilogram
1,000 milliliters = 1 liter

C. Choose the units you would use from the English and metric systems to measure each of the following items.

10.

_____

_____

11.

_____

_____

12.

_____

_____

To convert smaller metric units to larger units, reverse the process.

▶ To change a smaller metric unit to a larger unit, *divide by the number of smaller units* in the larger unit.

Example: 2,500 g = _____ kg
Think: 2,500 ÷ 1,000 = 2,500 = **2.5 kg**
Move decimal point 3 places.

D. Change the following measurements to the units indicated.

13. 500 mg = _____ g

14. 1,200 ml = _____ liters

15. 22 ml = _____ liter

16. 4,500 g = _____ kg

17. 800 g = _____ kg

18. 25 mg = _____ g

Answers start on page 241.

Measuring Temperature

Temperature measures the hotness or coldness of an object or environment. To measure temperature we use a **thermometer.** The most common thermometer used to measure air temperature is the weather thermometer.

The Weather Thermometer

The weather thermometer can contain two scales, the **Fahrenheit** scale (°F) and the metric **Celsius** scale (°C). Notice that each scale has both positive and negative numbers.

The temperature is indicated by the top of the liquid in the center tube. The temperature shown here is 68°F, which is equal to 20°C.

Each mark on the Fahrenheit scale is 2° more than the mark below it. Each mark on the Celsius scale is 1° more than the mark below it.

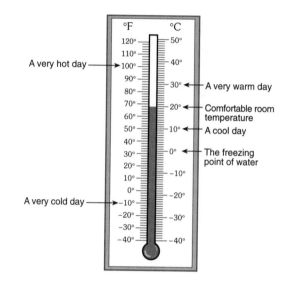

A. Review the temperature readings marked on the thermometer above. Based on these readings and your own experiences, choose appropriate Fahrenheit and Celsius temperatures for each of the following descriptions.

1. A very cold winter day.

   _____ °F

   _____ °C

2. An early spring day in New York.

   _____ °F

   _____ °C

3. Temperature at which puddles of water turn to ice.

   _____ °F

   _____ °C

4. A hot summer day.

   _____ °F

   _____ °C

5. A pleasant indoor temperature.

   _____ °F

   _____ °C

6. A very hot day.

   _____ °F

   _____ °C

B. Write each temperature reading in both °F and °C.

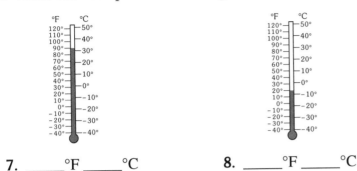

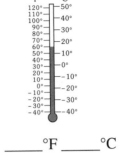

7. _____°F _____°C   8. _____°F _____°C   9. _____°F _____°C

C. Solve the following word problems using any of the thermometers pictured above.

10. During the evening the temperature in Bellvue dropped 8°F. If the temperature was 67°F, what did the temperature drop to?

11. The average August temperature in San Diego is 78°F. For the same month, the average temperature in New Orleans is 90°F. How much warmer is New Orleans in °C?

12. The temperature inside a tropical greenhouse was 90°F. Outside it was 10°C. How much warmer was it inside the greenhouse in °F?

13. On the thermometers pictured above, what is the difference between the highest and lowest readings on the Fahrenheit scale? On the Celsius scale?

Making Connections: The Clinical Thermometer

A second type of thermometer found in many homes is the **clinical thermometer.** The clinical thermometer is used to measure body temperature. The temperature range is narrow, limited to about 92°F to 110°F, or about 33°C to 44°C.

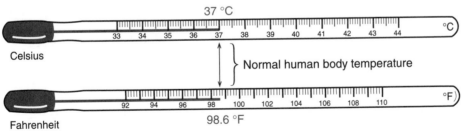

The thermometers pictured above show normal human body temperature as 37°C or 98.6°F.

A person's temperature can be taken by placing the bulb of a thermometer under the person's arm. This method always results in a low reading. For example, a normal temperature taken this way reads about 97°F.

Your child has a cold. Taken under her arm, her temperature is 99°F. What would you conclude about your child's temperature?

Reading Scales and Meters

The Pound Scale

To measure weight of a few ounces or pounds, use a pound scale like the one shown below. Each numbered division represents 1 lb., while the smaller divisions represent ounces or fractions of a pound.

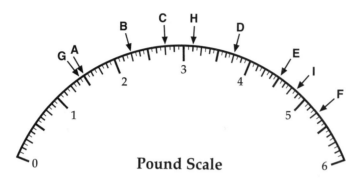

Pound Scale

You read a weight on this scale as pounds and ounces or as a mixed number of pounds. The largest markings on the scale indicate whole numbers of pounds. The large divisions between the pounds represent $\frac{1}{2}$ lb. or 8 oz. The smaller divisions show $\frac{1}{4}$ lb. or 4 oz. The smallest markings are 1 oz.

A. Using the pound scale above, write the weight represented by letters *A* through *I* as pounds and ounces and as a mixed number of pounds, where space is provided.

1. Point A = ___ lb. ___ oz.   Point B = ___ lb. ___ oz.   Point C = ___ lb. ___ oz.

_____ lb.   _____ lb.   _____ lb.

2. Point D = ___ lb. ___ oz.   Point E = ___ lb. ___ oz.   Point F = ___ lb. ___ oz.

_____ lb.   _____ lb.   _____ lb.

3. Point G = ___ lb. ___ oz.   Point H = ___ lb. ___ oz.   Point I = ___ lb. ___ oz.

B. Plot each of these weights on the scale above.

4. Point J = $1\frac{3}{4}$ lb.   Point K = 4 lb. 6 oz.   Point L = 3 oz.

Speedometers and Odometers

In a car, speed is indicated on a speedometer. The speedometer shown below has both English and metric speed readings. A car's speed is indicated by the position of the needle.

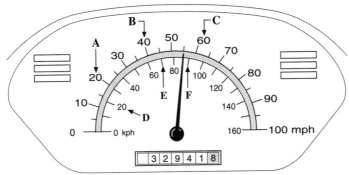

The English side is marked in *miles per hour* (MPH). This is the number of miles that the car will travel in one hour if it maintains that speed. The metric side measures the same speed in *kilometers per hour* (KPH).

C. **Using the speedometer above, write the speed represented by letters** *A* **through** *F* **as miles per hour (MPH) and as kilometers per hour (KPH).** (*Hint:* **When the needle does not point to an exact marking, give an approximate answer.)**

5. Point A = _____ MPH   Point B = _____ MPH   Point C = _____ MPH

     _____ KPH   _____ KPH   _____ KPH

6. Point D = _____ MPH   Point E = _____ MPH   Point F = _____ MPH

     _____ KPH   _____ KPH   _____ KPH

Near the bottom of the speedometer is the **odometer,** a device that keeps track of the total number of miles the car has traveled. (The last digit on the odometer indicates tenths of a mile.)

D. **Use the odometer pictured above to answer the following questions.**

7. What is the total number of miles shown on the odometer?

8. **List** Mention several ways to use the information provided by the odometer and speedometer.

Figuring Distance, Rate, and Time

Distance, rate, and time are all related. If you travel faster, you can cover more distance in less time. You can travel 200 miles in a lot less time if you drive at 50 miles per hour than if you drive at 20 miles per hour.

▶ The **distance formula** shows the relationship between distance, rate, and time: distance (d) = rate (r) $\times$ time (t), or $d = rt$.

If you know any two of the values in the distance formula, you can find the third by using one of the formulas shown in the following examples.

Using the Distance Formula

Example: Rudy drove an average of 50 miles per hour for $3\frac{1}{2}$ hours. How far did he drive?

Step 1
Write the distance formula to find the distance.

$d = rt$

Step 2
Substitute the known values for rate and time.

$d = 50 \times 3\frac{1}{2}$

Step 3
Multiply to find the distance.

$\frac{\overset{25}{\cancel{50}}}{1} \times \frac{7}{\underset{1}{\cancel{2}}} = 175$ miles

Rudy drove **175 miles** in $3\frac{1}{2}$ hours at 50 miles per hour.

A. **Find the distance traveled in the problems below. Round to the nearest mile.**

1. rate = 45 MPH
   time = 4 hours
   distance = _____ miles

   rate = 36 MPH
   time = $2\frac{1}{2}$ hours
   distance = _____ miles

   rate = 55 MPH
   time = $1\frac{3}{4}$ hours
   distance = _____ miles

   rate = 48 MPH
   time = $5\frac{1}{3}$ hours
   distance = _____ miles

Using the Rate Formula

Example: Theresa made the trip to her mother's house in $2\frac{1}{2}$ hours. If the distance she drove is 100 miles, what was her average speed (rate)?

Step 1
Rewrite the distance formula to find the rate.

$\frac{d}{t} = r$

Step 2
Substitute the known values for distance and time.

$r = 100 \div 2\frac{1}{2}$

Step 3
Divide to find the rate.

$\frac{\overset{20}{\cancel{100}}}{1} \times \frac{2}{\underset{1}{\cancel{5}}} = 40$ miles per hour

Theresa drove at an average speed of **40 miles per hour.**

B. **Find the average speed (rate) in the problems below. Round to the nearest mile per hour.**

2. distance = 120 miles
   time = 3 hours
   rate = ____ MPH

   distance = 152 miles
   time = $3\frac{1}{2}$ hours
   rate = ____ MPH

   distance = 276 miles
   time = $4\frac{3}{4}$ hours
   rate = ____ MPH

   distance = 598 miles
   time = 12 hours
   rate = ____ MPH

Using the Time Formula

Example: How long will it take Henri to travel 120 miles if he drives at an average speed of 40 miles per hour?

Step 1
Rewrite the distance formula to find the time.

$$\frac{d}{r} = t$$

Step 2
Substitute the known values for distance and rate.

$$t = 120 \div 40$$

Step 3
Divide to find the time.

$$40\overline{)120} = 3 \text{ hours}$$ with quotient 3

It will take Henri **3 hours** to travel 120 miles at 40 miles per hour.

C. **Find the time in the problems below. Round to the nearest hour.**

3. distance = 140 miles
   rate = 35 MPH
   time = ____ hours

   distance = 245 miles
   rate = 40 MPH
   time = ____ hours

   distance = 500 miles
   rate = 55 MPH
   time = ____ hours

   distance = 367 miles
   rate = 32 MPH
   time = ____ hours

D. **Solve these distance, rate, and time problems. Round your answer to the nearest whole number, if necessary.**

4. Wendy drove for 6 hours to visit her friends. Her average speed was 48 miles per hour. How many miles did Wendy travel?

5. The Backmans moved to a town 1,575 miles from their old home. If it took the moving company 36 hours to get to their new home, what was the average speed of the moving van?

6. Jan can drive to his brother's at an average speed of 32 miles per hour. The place is 30 miles away. How long does it take for Jan to get to his brother's house?

7. **Multiple Answers** Yosef drives 45 miles round-trip to and from work each day. How long does it take him (to the nearest *tenth* of an hour) to drive round-trip if he averages 25, 30, 35, 40, or 45 miles per hour depending on traffic?

Mixed Review

A. Change each measurement to the unit indicated.

1. $3\frac{1}{2}$ ft. = _____ in.   34 in. = _____ ft. _____ in.   5 yd. = _____ ft.

2. 17 ft. = _____ yd. _____ ft. $1\frac{1}{4}$ mi. = _____ ft.   12 fl. oz. = ___ c. ___ fl. oz.

3. $2\frac{3}{4}$ c. = _____ fl. oz.   46 oz. = _____ lb. _____ oz. $2\frac{2}{5}$ t. = _____ lb.

4. $1\frac{1}{8}$ gal. = _____ fl. oz.   18 oz. = _____ lb. _____ oz. $3\frac{1}{2}$ lb. = _____ oz.

5. 4.2 km = _____ m   17.1 g = _____ mg   2,750 ml = _____ liters

B. Add or subtract the following measurements.

6.
      3 ft. 8 in.   8 ft. 9 in.   2 yd. 2 ft.
   + 4 ft. 5 in.   − 5 ft. 6 in.   + 3 yd. 1 ft.
   _____   _____   _____

7.
      3 c. 6 fl. oz.   5 c. 1 fl. oz.   7 gal. 1 qt.
   + 2 c. 3 fl. oz.   − 4 c. 7 fl. oz.   − 3 gal. 2 qt.
   _____   _____   _____

8.
      4 gal. 3 qt.   2 lb. 6 oz.   4 lb. 8 oz.
   + 2 gal. 2 qt.   + 5 lb. 14 oz.   − 3 lb. 9 oz.
   _____   _____   _____

C. Solve the following measurement problems.

9. Ali has two jars of honey, shown below. How much honey is in each of the two jars? How much more honey is in the first jar? How much honey is there combined?

10. Kathryn is building a door frame. She cut the piece of 2 × 4 shown below. The piece is too long. What is the length of the wood? How long will it be after Kathryn cuts off $3\frac{7}{8}$ in.?

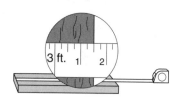

D. Solve the following word problems.

11. T.J. needs to buy drapes for his picture windows. The drapes must be $6\frac{1}{4}$ ft. long. The store measures the drapes in inches. How many inches long must T.J.'s drapes be?

12. Kevin rented a truck to move his family belongings. The truck weighs 2 t. 500 lb. empty. Kevin loaded about 800 lb. into the truck. How much does the truck weigh now that it is loaded?

13. Claudio had 2 lb. of chopped meat in the refrigerator. He needs 32 oz. of meat to make chili for the company picnic. Does Claudio have enough chopped meat?

14. Sarah had a piece of yarn measuring 5 yd. She used 2 yd. 2 ft. to decorate a craft project. How much yarn does Sarah have left?

15. Brenda drives 360 miles each week. She drives at an average of 35 miles per hour. How much time does Brenda spend driving each week? Round to the nearest hour.

16. What was Juliette's average speed in miles per hour if she drove 55 miles in 1 hour?

17. Antonia is cooking a holiday meal for her family. While preparing her shopping list, she finds that she needs 1 c. 3 fl. oz. of chicken broth in one dish and 2 c. 6 fl. oz. of broth in another. How much broth does Antonia need altogether?

18. The elevator in Assam's Department Store can hold 1 t. 200 lb. People and goods weighing 750 lb. are in the elevator. How much more weight can the elevator hold?

19. Bevin ran around a circular track 8 times. The distance around the track is 440 yd. How far did Bevin run in *miles?*

20. Ron was quarterback for his college football team. In four years he passed for a total of 7,208 yd. How many *feet* did Ron pass for in his career?

21. Francisco drove 55 kilometers per hour for 4 hours to reach an important conference. How far did he travel?

22. Elizabeth had to mix 3 fl. oz. of water with powdered milk. She measured the water with a *teaspoon*. How many teaspoons of water did Elizabeth need?

E. Answer these questions by using the thermometer shown.

23. What does the thermometer read in °F? in °C?

24. Which word would you use to describe this temperature: cold, cool, warm, or hot?

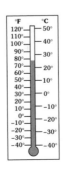

Tables and Charts

A **chart** is a representation of facts shown as numbers, words, or symbols. Maps, lists, and tables are forms of charts.

A **table** is often used to organize and display data, especially data that can be grouped into categories. A table may contain both words and numbers written in labeled rows and columns.

A **row** is read from left to right (or right to left).

| 125 387 87 299 |

read across ———→

A **column** is read from top to bottom (or bottom to top).

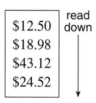

read down

Look at the population table below. The table contains a **title,** row and column **headings** (labels), and data.

Population of Top 5 U.S. Cities: 1980 and 1990
(in thousands)

City	1980 Population	1990 Population	1990 Rank	Percent Change 1980–1990
Chicago, IL	3,005	2,784	3	−7.4
Houston, TX	1,595	1,631	4	2.2
Los Angeles, CA	2,969	3,485	2	17.4
New York, NY	7,072	7,322	1	3.5
Philadelphia, PA	1,688	1,586	5	−6.1

Source: World Almanac Book of Facts, 1994

Finding Information in a Table

Example: What was the population of Los Angeles in 1990?

Step 1
Find the row with the desired row heading.
"Los Angeles, CA"

Step 2
Find the column with the desired column heading.
"1990 Population"

Step 3
Read the value at the point where the row and column meet.
3,485 thousand*

The population of Los Angeles in 1990 was 3 million 485 thousand, or **3,485,000.**

*The title explains that all population figures are in thousands.

A. Use the population table on page 134 to answer these questions.

1. What was the population of Chicago in 1980?

2. What rank did Houston hold in 1990?

3. Which city had the highest percentage of population growth?

4. Which city lost 6.1% of its population between 1980 and 1990?

B. Use the table below to answer questions 5–8.

Johan's Service Center Payroll

Employee	Hourly Pay Rate	Hours Worked	Gross Pay
Perez, Antonio	$12.50	40	$500.00
Hastings, William	$9.85	40	$394.00
Russel, Iris	$14.75	38	$560.50
Yamomoto, Mikki	$18.50	40	$740.00

5. Which employee has the highest hourly wage?

6. Which employee worked the *fewest* hours?

7. What was Iris Russel's gross pay?

8. Which employee earns $9.85 per hour?

C. Use the mileage chart to answer questions 9–12.

Airline Mileage to North American Cities

Cities	Mexico City	Montreal	New York	San Francisco
Mexico City		2312	2085	1883
Montreal	2312		330	2537
New York	2085	330		2566
San Francisco	1883	2537	2566	

Source: World Almanac Book of Facts, 1994

9. What is the distance between New York and San Francisco? (*Hint:* Using a row and a column, find where the two cities intersect.)

10. How much farther is it from Montreal to San Francisco than from Montreal to Mexico City?

11. How many miles will Armando travel if he flies from Mexico City to San Francisco and continues on to New York?

12. **Explain** Why are there four empty boxes on the chart?

Computer Spreadsheets

To complete many tables, you have to do quite a few calculations. Even if you use a calculator to do the calculations, they can take a lot of time.

This job can be made much easier if you use a computer **spreadsheet program.** A spreadsheet program uses a special table called a **computer spreadsheet** that automatically does calculations for you.

A computer spreadsheet looks like the table below. On a spreadsheet, each column is given a letter and each row has a number. The data goes inside rectangular boxes called **cells.** Each cell is identified by its column letter and row number, called the **cell address.**

Computer Spreadsheet

text (cell address D1)

	A	B	C	D	E
1				May	% Increase
2				$4,321.97	12.4%
3					

number (cell address E2)

A cell may contain any one of three types of information, or data:

text: Nonnumeric information is called *text* or *labels.* Words, symbols, years, and so on are examples of text. Text may be used to label column and row headings.	*numbers:* Numbers, including monetary amounts, are the most common forms of data on a spreadsheet.	*formulas:* A formula tells the computer to enter the results of a calculation in a cell. It consists of some combination of cell addresses, numbers, and operation signs (+, -, *, /).

Below are examples of spreadsheet formulas using numbers.
(*Note:* The sign * means multiplication; / means division.)

Addition	Subtraction	Multiplication	Division
=45+87	=132-78	=54*12	=55/19

Below are examples of formulas using cell addresses:

Addition	Subtraction	Multiplication	Division
=B1+A2	=E6-E7	=D4*F1	=C3/C12

A. Use the partially completed computer spreadsheet to solve problems 1–4.

	A	B	C	D
1	Name	Hours Worked	Pay Rate	Gross Pay
2	Johnson, Tim	40	$8.45	=B2*C2
3	Garcia, Maria	36	$9.50	

1. What is the gross pay amount that would be displayed in cell D2 if the formula was calculated?

2. What formula should be entered in cell D3 to determine Maria Garcia's gross pay?

3. What is the answer that would be displayed in cell D3 if the formula was calculated?

4. What formula would add the two gross pays and display the sum?

B. Use the spreadsheet below to solve problems 5–9.

	A	B	C	D
1	Item	Regular Price	Discount	Discount Price
2	Dress	$37.50		
3	Pants	$28.80		

5. Write the formula for cell C2 and the result the computer would calculate if the store is offering a 20% discount. (*Hint:* 20% = .2)

6. Write the formula for cell C3 and the result if the store is offering a 25% discount.

7. In cell D2 you want to subtract the discount in cell C2 from the regular price in cell B2. Write the formula and the result.

8. Write the formula and the result for cell D3 that subtracts the discount in cell C3 from the regular price in cell B3.

9. **Chart** Use the information in the chart below to complete rows 4, 5, and 6 of the spreadsheet. Include both the formula and the answer that would be displayed.

Item	Regular Price	Discount
Shoes	$37.90	30%
Shirt	$25.00	15%
Sunglasses	$18.50	50%

	A	B	C	D
4				
5				
6				

Bar Graphs

Graphs give a visual picture of data. In a **bar graph,** data is represented by the height (or length) of **data bars.** The bars may be drawn vertically (up or down) or horizontally (across).

The bar graph below has a **title** and labels on both **axes**—lines that show the possible values of the things being compared. The **vertical axis** shows the population in thousands of the five largest U.S. cities. **The horizontal axis** shows the names of the cities. Data is represented by the height of the vertical bars and their values on the vertical axis.

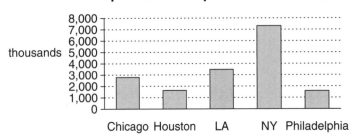

Population of Top 5 U.S. Cities 1990

Source: World Almanac Book of Facts, 1994

Finding Information on a Bar Graph

Example 1: What was the *approximate* population of Philadelphia in 1990?

Step 1
On one axis, locate the data you wish to look up.

Philadelphia is the fifth city listed on the horizontal axis.

Step 2
Read the value on the remaining axis closest to the height or length of the bar.

The bar is halfway between 1,000 and 2,000—about 1,500. Since the values are in thousands (see the label on the vertical axis), Philadelphia's population in 1990 was about **1,500,000.**

Example 2: Which city had the largest population in 1990?

Step 1
Find the highest or longest bar.

The fourth bar is the highest.

Step 2
Read the value on the horizontal axis at the bottom of the bar.

The bar represents New York's population. Thus **New York** had the largest population in 1990.

A. Use the bar graph on page 138 to solve the following problems.

1. Which city had the smallest population in 1990?

2. Which city had an approximate population of 3,500,000 in 1990?

3. *Estimate* the population of Chicago in 1990.

4. How accurately do you think *you* can read data on this bar graph?

The bar graph below has two horizontal bars for each city. A **legend** explains the symbols used in a graph. The legend at the right shows that the top bar represents 1980 population and the bottom bar 1990 population.

Population of Top 5 U.S. Cities 1980 and 1990

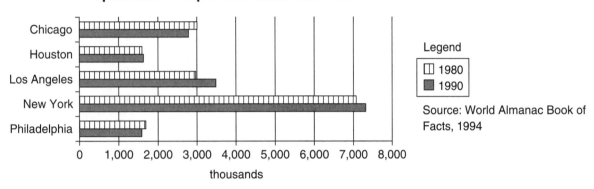

B. Problems 5–8 refer to the double bar graph above.

5. Which cities show an *increase* in population between 1980 and 1990?

6. Which cities show a *decrease* in population between 1980 and 1990?

7. Which city shows the *largest increase* in population between 1980 and 1990?

8. Which city shows the *smallest increase* in population between 1980 and 1990?

Making Connections: Drawing a Bar Graph

Use the axes at right to create a vertical double bar graph of monthly household expenses in 1994 and 1995.

Expense	1994	1995
Rent	$495	$510
Food	$325	$350
Utilities	$150	$125

Line Graphs

A **line graph** displays data as positions along a line, drawn from left to right across the page. The value of each **data point** is found by using a value from each axis. In a line graph, each axis contains a scale of values. Line graphs are especially useful when showing how a quantity changes over time.

The line graph at the right shows the **gross national product** of the United States from 1960 to 1990. The gross national product is the total value of goods and services produced in a nation. The dots represent the data points. From the graph it is easy to see that the gross national product is increasing over time.

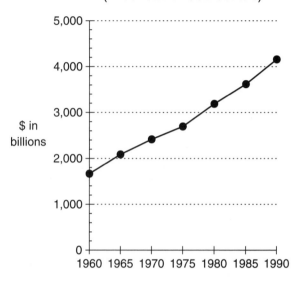

U.S. Gross National Product
(in constant 1982 dollars)

$ in billions

Source: U.S. Bureau of Economic Analysis

Finding Information on a Line Graph

Example: What was the *approximate* gross national product of the United States in 1970?

Step 1
Locate 1970 on the horizontal axis and find the data point directly above it.

1970 is the third data point from the left.

Step 2
Read the value on the vertical axis closest to the data point.

The data point is just below $2,500. You might estimate that in 1970 the gross national product was about $2,500 billion, or **$2,500,000,000,000.**

A. Use the line graph above to solve the following problems.

1. What was the approximate gross national product in 1965?

2. In which year was the gross national product about $3,000 billion?

3. Which 5-year period saw the *smallest* increase in gross national product?

4. Which period saw the *largest* increase in gross national product?

B. Use the line graph below to solve problems 5–8. This line graph contains three lines, each representing one set of data. Use the legend to find out which line goes with each country.

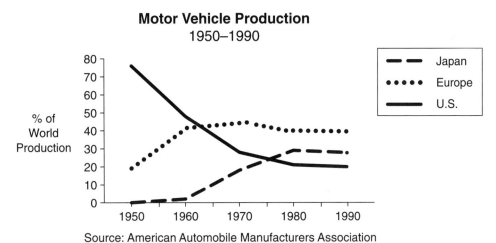

Motor Vehicle Production
1950–1990

Source: American Automobile Manufacturers Association

5. Which producer (country or region) had the largest share of the market in 1950?

6. Estimate in which year Europe first produced more cars than the United States.

7. Estimate in which year Japan first produced more cars than the United States.

8. In which 10-year period did Japan show its *largest* increase in its share of car production?

C. Use the line graph below to solve problems 9–12.

Note: If you listed the age of each person in the United States from youngest to oldest, the median age would be the age of the person in the middle of the list.

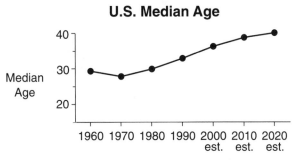

U.S. Median Age

Source: U.S. Bureau of the Census

9. What was the median age of the U.S. population in 1980?

10. In which period did the median age drop?

11. In which period is the increase in median age *greatest?*

12. **Describe** Based on your knowledge of U.S. history and culture, what are some factors that could explain the rise in median age?

Circle Graphs

A **circle graph** displays data as sections of a circle. Circle graphs are especially useful to show how a whole amount such as a budget is made up of several parts.

In a circle graph the circle represents the whole amount. Each segment represents a part of the whole. The segments may be labeled to describe the part or they may be shaded or colored. In that case, a legend relates the color or shading to the description.

Data on the circle can be displayed as a percent, a fraction, or cents per dollar. The whole circle can equal 100%, 1, or $1.00.

The circle graph pictured below shows motor vehicle production around the world.

Motor Vehicle Production 1992

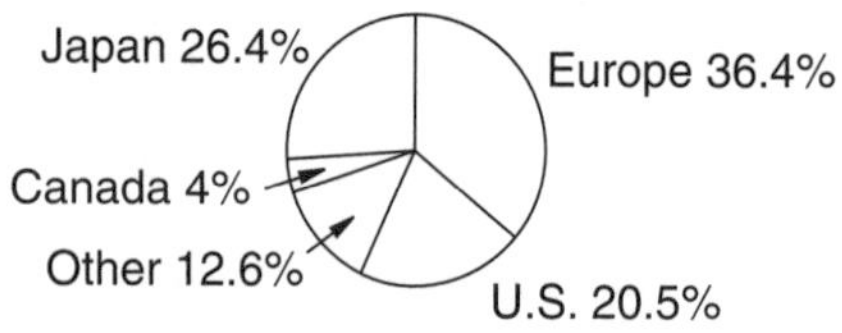

Japan 26.4%   Europe 36.4%

Canada 4%

Other 12.6%   U.S. 20.5%

Source: American Automobile
Manufacturers Association

Finding Information on a Circle Graph

Example: Which country or region produced the most vehicles in 1992?

Step 1
Find the largest segment in the circle and read the label.

Step 2
Read the value in the segment.

The largest segment is labeled "Europe."

Europe—36.4%

A. Use the circle graph above to solve the following problems.

1. What percent of vehicles made in 1992 were produced by Japan?

2. What percent of vehicles made in 1992 were produced by the United States *and* Canada?

3. What percent of vehicles made in 1992 were *not* produced by the United States?

4. The Canadian share was 3.4% in 1980. Did its percentage increase or decrease between 1980 and 1992?

B. Use the circle graph below to solve the following problems.

Copper Use in the U.S.

Source: Bureau of Mines, U.S. Department of the Interior

5. What is the greatest use of copper in the United States?

6. What percent of copper is used in transportation?

7. What percent of copper is used in either machinery or consumer products?

8. What percent of copper is *not* used in building construction?

C. Use the circle graph below to solve the following problems.

Sanchez Company Workforce

9. What is the largest group of employees in the company?

10. What percent of the company employees are classified as Support?

11. What percent of the company employees are involved in either Sales *or* Manufacturing?

12. If the total number of employees is 265, how many people are in Management?

Making Connections: Creating a Circle Graph

The Williams family's monthly expenses are listed below. Add the amounts to find the total expenses. Then use the total to find the percent for each expense. Write the percents in the circle graph on the right. For example, to figure the food expense, divide $600 by the total expenses and write the answer as a percent.

Housing $850
Food $600
Transportation . . .$150

Medical Expenses . . .$175
Other $225

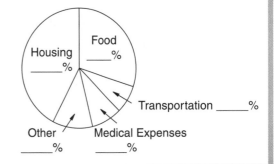

Williams Family Budget

Scatter Diagrams

A **scatter diagram** (also known as a *scattergram*) is a graph that shows a possible relationship between two different groups of data.

The table shows partial results of a survey. People were asked the highest level of school they completed and their yearly income.

Name	Schooling	Income
Destafano	high school	$18,700
N'komo	bachelor's degree	$32,600
Viramentes	some college	$20,900

The results of the survey are represented in the scatter diagram below. Each *X* on the scatter diagram represents one person. The position of the *X* on the horizontal axis represents the level of school completed. The height of the *X* shows the income.

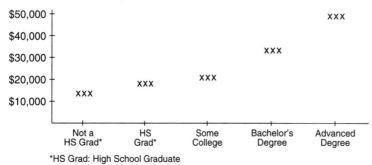

Yearly Income by Level of Education

*HS Grad: High School Graduate

Reading a Scatter Diagram

Example: Carol is a junior in high school. She wonders whether it is worthwhile for her to finish high school. According to the scatter diagram, how much more will a high school graduate earn per year than someone who doesn't finish school?

Step 1
Locate the values or labels on the horizontal axis to be checked.

"Not a HS Grad" and "HS Grad" are the categories to be checked.

Step 2
Estimate the average value for each category. Find the difference.

"Not a HS Grad" ≈ $13,000
"HS Grad" ≈ $18,000

$18,000 − $13,000 = $5,000

A high school graduate earns about **$5,000** more per year than someone who doesn't finish high school.

A. Use the scatter diagram on page 144 to solve problems 1–4.

1. Estimate the average salary earned by a person with an advanced degree.

2. Estimate the average salary earned by a person with some college but no degree.

3. Estimate the difference in salary between someone with a high school diploma and someone with a bachelor's degree.

4. What do the results of the survey say about the relationship between education and income?

B. Use the data in the following table to complete the scatter diagram at right.

Driving Time in Minutes

	M	T	W	T	F
Don	26	24	21	25	23
Miguel	18	21	17	18	20
Elsa	31	29	26	28	30

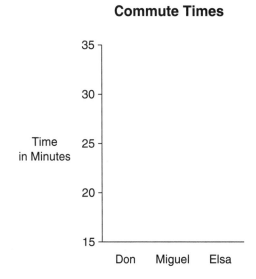

Making Connections: Choosing an Appropriate Graph

Each type of graph has features that make it especially useful for certain types of data.

For each description of data below, tell which type of graph you think would be best suited to show the data. Choose from bar, line, and circle graphs and scatter diagrams. Give a reason for your answer.

1. The presidential election returns showing the percent of the vote for each of the three leading candidates.

2. A survey of children that shows a relationship between the number of hours they watch television and their school grades.

3. The average temperature for each month of the year in two cities in the same state.

4. A comparison of acres of different types of grain planted in a county.

Using More than One Data Source

Sometimes information from more than one source of data is needed to answer a question. Tables, graphs, and text may contain only one part of the information you need.

Suppose Greg is planning to vacation in California. He will be driving from Los Angeles to San Francisco, and he wants to rent a car. Greg knows that gasoline costs about $1.20 per gallon. What will be the difference in gasoline cost of driving the various rental cars?

The two charts below each contain information that can be used to answer the question. The mileage chart will let Greg find the distance between the two cities. The bar graph shows the number of miles per gallon each rental car gets.

	Atl.	Bos.	Chi.	Dal.	LA	NO	SF
Atl.		1037	674	795	2182	479	2496
Bos.	1037		963	1748	2979	1507	3095
Chi.	674	963		917	2054	912	2142
Dal.	795	1748	917		1387	496	1753
LA	2182	2979	2054	1387		1883	379
NO	479	1507	912	496	1883		2249
SF	2496	3095	2142	1753	379	2249	

Source: Rand McNally Road Atlas, 1994

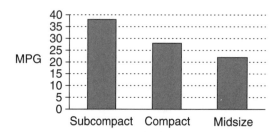

Using More than One Source

Example: How much will it cost Greg for gas to drive from Los Angeles to San Francisco if he rents a compact car and the price of gasoline is $1.20 per gallon?

Step 1	**Step 2**	**Step 3**
Set up the problem.	Find the values in the tables, graphs, and/or text.	Do the arithmetic.
mi. ÷ mpg = gal. gal. × price per gal. = cost	mi. = 379 (mileage chart) mpg = 28 (bar graph) price per gallon = $1.20 (text)	379 ÷ 28 = 13.5 gal. (rounded to the nearest tenth) 13.5 × $1.20 = $16.20

Greg would spend **$16.20** on gas if he rented the compact car.

A. Solve problems 1–4 using the table and graph on page 146. You may use a calculator.

1. What would it cost for gas to drive the midsize car from Boston to Chicago if gas costs $1.15 per gallon? (Round the gallons to the nearest tenth and the cost to the nearest cent, if necessary.)

2. How many gallons would it take to drive from New Orleans to Atlanta in the subcompact? (Round to the nearest gallon.)

3. How much more would it cost for gas to drive a compact car than a subcompact from Dallas to New Orleans if gas cost $1.20 per gallon?

4. Assume the grade of gas for the compact car costs $1.10 and the grade of gas for the midsize costs $1.25. How much more would gas for the midsize cost on a trip from San Francisco to Chicago?

B. Use the graphs below to solve the following problems.

1992 Colorado Presidential Election Results

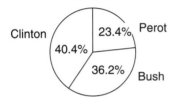

Source: World Almanac Book of Facts, 1995

1992 U.S. Presidential Election Results

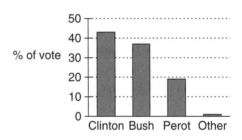

Source: World Almanac Book of Facts, 1995

5. If a total of 1,561,541 people voted in Colorado, how many people voted for Bush?

6. If 104,552,736 people voted in the United States, about how many voted for Clinton?

7. Did Perot get a larger or smaller percent of the vote in Colorado than in the entire country?

8. How do the percent of the votes for Clinton and Bush compare between Colorado and the United States as a whole?

C. Use the circle graph above and the graph below to solve problems 9 and 10.

9. Did Clinton get a higher or lower percent of the vote in Adams County than in the state vote?

10. **Compare** In which counties did Bush receive a higher percent of the vote than in the state vote?

1992 Colorado Presidential Election Results

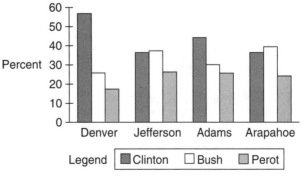

Source: World Almanac Book of Facts, 1995

Simple and Compound Probability

Probability is the likelihood (or chance) of something happening or not happening. Probability, the study of chance, is a useful data analysis tool.

The probability that an event will occur (called a **favorable outcome**) depends on two numbers: the number of ways the favorable outcome can occur and the total number of possible outcomes (favorable and unfavorable).

Probability of an event $= \frac{\text{number of favorable outcomes}}{\text{total number of possible outcomes}}$

▶ A probability of 0% means that an event cannot happen.

▶ A probability of 100% means that an event will happen for certain.

In determining **simple probability** you must find the chance of a single event occurring.

Finding Simple Probability

Example: Oscar is having a hard time deciding which socks to wear. He has 8 pairs of socks, folded together, in his drawer. Two of the pairs are blue. If Oscar just reaches in and chooses a folded pair without looking, what are the odds that he will choose a blue pair?

Step 1	**Step 2**	**Step 3**
Find the total number of outcomes.	Find the number of favorable outcomes.	Write the favorable number as the numerator and the total number as the denominator. Simplify if possible.
Number of pairs of socks $= 8$	Number of blue pairs $= 2$	$\frac{\text{favorable}}{\text{unfavorable}}$ $\quad \frac{2}{8} = \frac{1}{4}$

The chance that Oscar will take a blue pair of socks is $\frac{1}{4}$. The answer can also be expressed as the percent **25%** or written as **1 chance in 4.**

A. Find the simple probability in the following problems.

1. The chance that you will draw a 7 out of 7 cards numbered 1 through 7.

2. The odds of rolling an odd number on a 6-sided die.

3. The chance that you will *not* roll a 6 on a 6-sided die.

B. The deck of cards below has 52 cards in all. There are 4 suits (diamonds, clubs, spades, and hearts) of 13 cards each. Each suit has an ace, king, queen, jack, and the numbers 2 through 10. Use this information to solve problems 4–7.

4. If a card is selected at random, what is the probability of choosing a spade?

5. What are the odds of selecting an ace at random?

6. What is the chance that you will randomly draw a 2 or a 3 of any suit? (*Hint:* Count the total number of 2s and 3s.)

7. What is the probability of drawing the ace, king, or queen of hearts from the deck?

In **compound probability** you have to find the chance of two or more events occurring. **Successive events** are events that happen one after another.

▶ To find the probability of successive events occurring, multiply the probability of the first event by the probability of the second event, and so on.

Finding Compound Probability

Example: Ella and Tom want to have two children. What is the chance that they will have two girls?

Step 1
Find the probability of the first event occurring.

$$\frac{1 \text{ (girl)}}{2 \text{ (girl or boy)}}$$

Step 2
Find the probability of the second event occurring, and so on.

$$\frac{1 \text{ (girl)}}{2 \text{ (girl or boy)}}$$

Step 3
Multiply the two probabilities.

$$\frac{1}{2} \times \frac{1}{2} = \frac{1}{4}$$

The odds of having two girls are $\frac{1}{4}$ or **25%** or **1 out of 4.**

C. Solve these compound probability problems.

8. What are the chances of flipping a coin 2 times and having heads come up both times?

9. What are the chances of flipping a coin 3 times and having tails come up all 3 times?

10. What are the odds of rolling two 6-sided dice and having a 4 come up on both dice?

11. **Explain** The odds of rolling a total of 7 with two dice is 1 out of 6. Why is this true?

Seeing Trends, Making Predictions

Probability can be based on the laws of chance. Probability can also be based on data. Graphs, tables, and other forms of data can be the basis of making **predictions.** Predictions based on available data give the best guess about future events. A prediction doesn't say what will happen, only what's most likely.

Basing Predictions on Data

Example: Of the students in a woodworking class, 62% are men, while 38% are women. What is the probability that the next student to enroll in the class will be a man?

Step 1
Find the category that corresponds to the event you are predicting.

The category to check is the percent of men in the class.

Step 2
The data in that category represents the likelihood that the event will occur in the future.

62% of the students are men.

You can predict that there is a **62%** probability that the next student to enroll in the class will be a man.

▶ To predict the number of times something will occur, multiply the probability of its happening once by the number of times it could occur, as in the following example.

Using Probability for Making Predictions

Example: If you roll two 6-sided dice 30 times, how many 7s will you probably roll?

Step 1
Determine the probability of rolling a 7 with 2 dice.

$$\frac{\text{favorable outcomes}}{\text{total possible outcomes}} = \frac{6}{36} = \frac{1}{6}$$

Step 2
Multiply the probability by the number of times it could occur.

$$\frac{1}{6} \times 30 = 5$$

The probability is that you will roll **5** sevens if you roll the dice 30 times.

A. Solve problems 1–4, basing your predictions on data and probability.

1. Bert has 3 blue shirts, 4 white shirts, and 1 yellow shirt. If he selects a shirt at random, what is the probability that Bert will wear a white shirt today?

2. Of all Orange Grove residents, 22% approve of their mayor, 24% don't approve, and the rest are undecided. What is the chance that Tony, a resident, is undecided?

3. Emanuel flipped a coin 10 times. How many times did the coin come up heads, probably?

4. State Contracting has won 7 of the 10 bids they made this year. Other things being equal, how many bids do you predict they will win out of the next 20?

Trends show how situations change over time. Using a trend, you can make an educated guess about what will occur, if the trend continues.

B. Solve problems 5–7 by using the trends shown on the line graph. The graph shows the amount of each foreign currency needed to equal one U.S. dollar over the years. As a currency increases in value, less of that currency is needed to equal a U.S. dollar.

Foreign Exchange Rates per U.S. Dollar

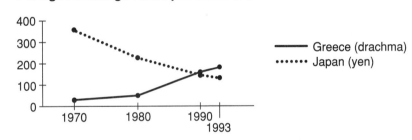

Source: World Almanac Book of Facts, 1995

5. What trend do you see in the value of the Japanese yen relative to the U.S. dollar?

6. What trend do you see in the value of the Greek drachma relative to the U.S. dollar?

7. If the trends continue, the value of which foreign currency do you predict will drop the most in the next year, compared to the U.S. dollar?

Making Connections: Checking Probability

The probability of rolling a number with two 6-sided dice is shown below.

2: $\frac{1}{36}$ 3: $\frac{1}{18}$ 4: $\frac{1}{12}$ 5: $\frac{1}{9}$ 6: $\frac{5}{36}$ 7: $\frac{1}{6}$

8: $\frac{5}{36}$ 9: $\frac{1}{9}$ 10: $\frac{1}{12}$ 11: $\frac{1}{18}$ 12: $\frac{1}{36}$

Predict the number of times you would roll a 6 if you rolled dice 18 times. Then, roll two 6-sided dice 18 times to see if your prediction comes close. Do you think your predictions would be more or less accurate if you rolled the dice more times?

Mean, Median, and Mode

A **typical value** is an amount that represents the values in a set of numbers. A typical value must fall within the series of numbers it represents. There are three common methods for finding typical values in a set of numbers: **mean, median,** and **mode.**

When most people use the word *average,* they are talking about the mean. For a set of values that are similar to each other, the mean is what you want to know when you ask, "About how much . . . ?"

To find the mean, add the values in a set of data, then divide the sum by the number of values in the set.

Finding the Mean

Example: Ella scored 14, 18, 11, and 16 points in her last 4 basketball games. What is her scoring average?

Step 1
Find the mean. Add the values in the set.

$$14 + 18 + 11 + 16 = 59$$

Ella's scoring average is **$14\frac{3}{4}$** points.

Step 2
Divide the sum by the number of values in the set.

$$59 \div 4 = 14\frac{3}{4}$$

▶ The mean is not necessarily equal to any number in the set. However, it must fall within the series of numbers in the set.

A. Find the mean in each of these sets of numbers. You may use a calculator.

1. 35, 49, 31, 25 $1.25, $2.08, $.97, $3.12, $1.45   5 lb., 4 lb., 12 lb.

2. 174, 12, 398, 214, 45   6 oz., 3 oz., 15 oz., 9 oz., 4 oz. $35, $12, $97, $6, $51, $88

The mean does not always represent a typical value very well, especially when one or more values are very large or small compared to the other values in the set. When this occurs, there are other methods of finding a more typical value.

The **median** is the middle value of a set of data. To find the median, sort the data from smallest to largest. For an odd number of values, the middle number is the median. For an even number of values, the median is the mean of the two middle numbers. The median is an accurate typical value when there are a few extreme values in a set of data.

Finding the Median

Example: Angela has helped 5 dogs onto the examination table. Their weights are 40 lb., 10 lb., 140 lb., 35 lb., and 30 lb. What is the median weight of the dogs?

Step 1
Sort the numbers numerically from smallest to largest.

10 lb., 30 lb., 35 lb., 40 lb., 140 lb.

Step 2
Choose the middle number from the set. (Find the mean of the two middle numbers, if necessary.)

35 lb. is the median weight.

B. Find the median in the following sets of numbers.

3. $18, $75, $4, $23, $51   53, 22, 12, 18, 50   13.4, 8.6, 4.1, 80.9

A third way to determine the typical value is by finding the **mode**—the value in the set that occurs most frequently. (If each value occurs only once, there is no mode in the set.) The mode is a good way to represent a set of values if one value occurs repeatedly.

Finding the Mode

Example: A restaurant offers these prices on complete meals: Fried Chicken—$8.99, Steak—$10.99, Fish—$8.99, Shrimp—$10.99, Lasagna—$8.99, Hamburger—$7.99. What is the mode for these prices?

Step 1
Sort the different values from least to greatest.

$7.99 $8.99 $10.99

Step 2
Count how many times each value occurs.

$7.99 $8.99 $10.99
  1   3   2

Step 3
Find the value that occurs the most.

$8.99 is the mode. It occurs the most (3 times).

C. Find the mode in each of these sets of numbers.

4. 35, 18, 21, 21, 18, 97, 32, 18   22°, 17°, 31°, 22°, 18°, 35°, 22°, 31°

D. Solve this problem.

5. Evaluate Melissa received the test scores at right. Which value—mean, median, or mode—would you use to describe Melissa's typical test score? Explain your choice.

Test Scores							
98	86	81	94	98	100	98	96

Unit 4 Review

A. Use the table below to solve the following problems.

Three-on-Three Basketball: Team A

Name	Age	Height	Weight
Davis	28	6 ft. 2 in.	190 lb.
Stern	41	6 ft. 4 in.	220 lb.
Norris	32	5 ft. 9 in.	175 lb.

1. Which member of the team is the shortest?

2. What is the *combined* weight of the team?

3. What is the approximate *average* age of the team?

4. **Discuss** Do you think Stern's age greatly affects the average age of the team? Compare the median age of the team to your answer in problem 3 to check your prediction.

B. Use the graph below to solve problems 5–10.

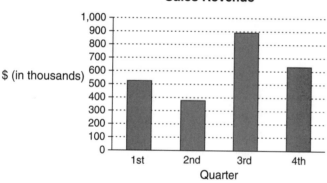

Osaki Manufacturing Company Sales Revenue

5. What was the approximate revenue of the company in the fourth quarter?

6. In which quarter was the sales revenue under $400,000?

7. Estimate the *mean* revenue for the first and second quarters.

8. In which quarter did the company have the *most* revenue?

9. What was the approximate total revenue for the company in the year? (*Hint:* The total revenue is the sum of the revenues for all four quarters.)

10. About how much *more* was the sales revenue in the third quarter than in the second quarter?

154

C. Use the circle graph and spreadsheet to solve problems 11–14.

Cole Family Expenses
($3,234 monthly expenses)

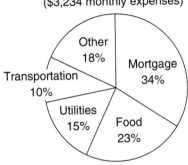

	A	B
1	Name	Income
2	Bill	$2,200
3	Mary	$1,300
4	Total	

11. What was the *largest* expense for the Cole family?

12. What formula would you enter into cell B4 of the spreadsheet if you wanted to calculate the total income?

13. How much does the Cole family spend on transportation each month?

14. What is the total income for the Cole family? How does their total income compare with their total expenses?

D. Solve these problems on probability, prediction, and trends.

15. What are the chances that someone you just met has a birthday in May if the person was not born in a leap year? What if the person *was* born in a leap year?

16. What are the odds that a coin flipped 3 times will have come up tails each time?

17. Valley High has won 5 of 8 games. At this rate, how many games would they win in the next 24 games?

18. What is the probability that a coin will land heads up when flipped?

19. A wheel has 20 sections, numbered 1 through 20. What are the odds that it will land on an *even* number when spun?

20. Bruce spends $659 per month on rent. Last year his rent was $629 per month. The year before that, the rent was $599. Based on this trend, how much do you think Bruce's rent will be *next* year?

Working Together

Conduct a survey in your area or class. The topic can be anything for which there are four or more categories or types of responses (suggestions: hobbies, pets, jobs, or music). Present the results in graph and table form. Make two statements that can be justified based on the results.

Algebra and Geometry

Skills

Order of operations

Using the distributive property

Writing expressions and equations

Powers and roots

Solving equations and inequalities

Understanding basic geometry figures

The Pythagorean theorem

Perimeter, area, and volume

Using the coordinate system

Tools

The number line

Protractors

Problem Solvers

Substituting to solve equations

Translating words to equations

Finding patterns in algebra and geometry

Choosing area, perimeter, or volume

Applications

Reading maps

Working with formulas

As we use math in our lives, the same types of problems come up again and again. Once we are familiar with a type of problem, we can solve similar problems by using what we have learned. **Algebra** allows us to write general rules to solve similar problems.

For example, you can find the cost of buying three $5 items by multiplying $5 by 3. Using algebra, we can write a **rule** for finding the cost of any number of items that have the same price.

Geometry is the study of points, lines, angles, and flat and solid figures. Everything built by people depends upon the principles of geometry.

In this unit, you will learn the basic principles of algebra and geometry. You will use these tools to solve problems.

When Do I Use Algebra and Geometry?

The following situations use algebra or geometry. Check the experiences you have had.

- ☐ figuring out which frame to buy for a picture
- ☐ estimating how long it will take you to drive a certain distance
- ☐ figuring out the amount you earn per hour, per day, or per week
- ☐ organizing the space in a closet
- ☐ estimating how much paint you need to paint a room

Describe how you would use algebra or geometry in these situations.

1. Tastyfood orange juice is priced at $1.68 for a 12-fluid-ounce can of concentrate. Using the same rate, how could you figure out how much a larger can would cost?

   _____

   _____

2. A state charges 8% sales tax. How could you figure out how much sales tax you owe on any purchase?

   _____

   _____

3. Have you ever rearranged furniture? Describe the process you used. Did you measure first, estimate sizes, draw diagrams, or just start moving the furniture?

   _____

   _____

4. On the job, you learn that to calculate the amount of simple interest a customer will owe, you need to multiply the principal (the amount borrowed) by the amount of time the money is borrowed for by the rate of interest. How could you write this idea using symbols to make it easier to remember?

   _____

   _____

Talk About It

A right angle is a square corner. Spend 5 minutes listing right angles you see in your classroom. Compare lists with a partner. Discuss why you think right angles are so common in artificial objects.

Throughout this unit, you may use a calculator wherever that would be useful.

Writing Expressions

An **expression** is a mathematical idea written with symbols. Expressions use the four operations of addition (+), subtraction (−), multiplication (×), and division (÷).

Addition and Subtraction

Addition and subtraction are easy to recognize. Addition combines amounts, and subtraction separates amounts.

<table>
<tr><td>

Addition

$4 + 6 = 10$

$\quad\quad\quad\llcorner$ sum

In addition, the order of the numbers being added doesn't matter.

Example: $1 + 4$ *is the same as* $4 + 1$

When you add 0 to a number, you get the original number.

Example: $8 + 0 = 8$

</td><td>

Subtraction

$10 − 4 = 6$

$\quad\quad\quad\llcorner$ difference

In subtraction, the order of the numbers *is* important.

Example: $9 − 4$ *is not the same as* $4 − 9$.

When you subtract 0 from a number, you get the original number.

Example: $7 − 0 = 7$

</td></tr>
</table>

Addition and subtraction are **inverse operations,** which means they are opposites of each other. If you add 3 to a number, you can undo the result by subtracting 3.

Example: $19 + 3 = 22$ and $22 − 3 = 19$

A. **Write an expression for each problem. Do not solve the problem. The first one is done as an example.**

$\underline{26 + 8}$ **1.** Ted's workweek of 26 hours increased by 8 hours.

_____ **2.** Nita spent $320 on food in March. That amount decreased by $45 in April.

_____ **3.** By Wednesday, Southern California's rainfall total was 15.6 inches. The region received another 2.8 inches of rainfall on Thursday.

_____ **4.** Cathy scored 12 points total in the first 3 quarters of the game and 8 points in the last quarter.

_____ **5.** The leather jacket that originally sold for $195 was marked down $50.

Multiplication and Division

To help you recognize when multiplication or division is needed, remember:
multiplication combines equal amounts, and division separates equal amounts.

Multiplication

$3 \times 6 = 18$

      └ product

In multiplication, the order of the numbers being multiplied doesn't matter.
Example: 3×6 *is the same as* 6×3.

Other ways to show multiplication are:

using a raised dot $6 \cdot 3$

using parentheses $6(3)$ or $(6)(3)$

Division

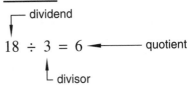

$18 \div 3 = 6$ ◄──── quotient

In division, the order of the numbers *is* important.
Example: $18 \div 6$ *is not the same as* $6 \div 18$.

Other ways to show division are:

using a division bracket $6\overline{)18}$

using a fraction bar $\frac{18}{6}$

using a slash $18/6$

Multiplication and division are inverse (opposite) operations.
If you multiply 5 by 6, you can undo the result by dividing by 6.
Example: $5(6) = 30$ and $30 \div 6 = 5$

B. Write two expressions for each problem using a different way to show the operation each time. Do not solve.

6. Mark's employer deducted $12 from his pay each month for 9 months.

7. Angela will split a $450 bonus evenly among her 3 employees.

8. If 21 classrooms each hold 30 students, how many students are there?

9. Five lab technicians equally shared the testing of 75 blood samples.

10. Fifteen people bought the vegetarian plate at $6.95.

C. A letter is used to represent the unknown amount of each block. Use the letters to write expressions for each problem below.

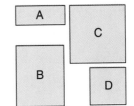

11. the sum of A and C

12. the difference between C and D

13. the product of A and B

14. **Multiple Solutions** List expressions showing C divided by D.

The Number Line

One way to show numbers and their relationship to each other is on a **number line.** The spaces on a number line represent regular intervals such as units (1, 2, 3, . . .) or tens (10, 20, 30, . . .) or fractions ($\frac{1}{4}, \frac{1}{2}, \frac{3}{4}, . . .$).

The number line on the right shows the whole numbers from 0 to 10. Whole numbers are also called **integers.** The arrows at the ends mean that the numbers continue in both directions.

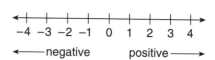

Values increase as you move to the right along the number line. What happens as you move to the left? They decrease.

In algebra we use **signed numbers.** Signed numbers are either **positive** (+) or **negative** (–). A number without a sign is understood to be positive. Both +4 and 4 are positive.

Where do negative numbers fit on the number line? A negative number has a value *less than 0,* while a positive number *is greater than 0.* Negative numbers are found to the *left* of 0 on a number line.

Using a Number Line to Solve Problems

Example: The temperature in Cheyenne was –10°. By midday, the temperature had increased 30°. What was the temperature at midday?

Step 1

The thermometer is a kind of number line. Find the starting temperature on the number line. Notice that positive numbers are above 0 and negative numbers are below.

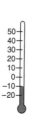

Step 2

Since the temperature increased, count up 30 degrees.

$-10° + 30° = 20°$

Numbers without a sign are positive.

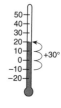

The temperature at midday was **20°** above 0.

You solved this problem by counting on the number line, but you were actually adding. To add a positive number, you move in a positive direction. To add a negative number, you move in a negative direction.

A. Add the signed numbers. Draw a number line if you need to.

1. $-5 + 15 =$ $+4 + 13 =$ $5 + (-18) =$ $-9 + 9 =$

2. $+10 + 12 =$ $12 + (-6) =$ $-2 + 2 =$ $-8 + 10 =$

Adding Two Negative Numbers

Example: Zoila owes her friend $20. Her debt can be expressed as –20. If she borrows another $10, how much will she owe?

Step 1. Write the expression.

–20 + (–10) =

Step 2. Use the number line below. To add –10, move to the left.

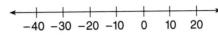

–20 + (–10) = –30

–30 means Zoila owes money. She owes **$30.**

B. Solve the following. Draw a number line if needed.

   3. –6 + (–28) =   –14 + (–13) =   –10 + (–5) + 20 =   –3 + 5 + –8 =

Subtracting a Negative Number

Example: At midnight, it was 2° in Fairbanks. Three hours later, it was 19° below zero. What was the difference in the temperatures?

Step 1. Write the expression.

2 – (–19) =

Step 2. Use the number line below. Count the distance between the two temperatures.

Hint: To subtract a negative number, change the negative sign to positive and add. 2 – (–19) = 2 + 19 = 21

There was a difference of **21°** between the two temperatures.

C. Solve the following signed number problems.

   4. 5 – (–12) =   –6 – (–2) =   7 – (+10) =   10 – (–25) =   –2 + 6 – (–3) =

Making Connections: Keeping Score

Robin	Art	Colin	Cassady
~~90~~	~~50~~	~~120~~	~~40~~
–70	90	220	–30

In many card games, players gain and lose points on each hand. Use the scorecard to answer the following.

1. How many points does Robin need to catch up to Art?

2. The most points a player can earn in one hand is 120. What is the smallest number of hands Colin would need to reach a winning score of 500?

3. On the current hand, Robin lost 50 points and Cassady gained 80 points. Colin lost 30 points and Art gained 60 points. Update the scorecard to show the changes in the players' scores.

Powers and Roots

In mathematics, a **power** shows repeated multiplication of a number by itself. The expression 5^2 means "5 to the second power." In this expression, 5 is the **base** and 2 is the **exponent**. To solve a power, multiply the base by itself the number of times shown by the exponent. $5^2 = 5 \cdot 5 = \mathbf{25}$

A number raised to the second power is called a **square**. $5^2 = 25$ is read as "five **squared** equals twenty-five." An expression showing a number raised to the third power such as $5^3 = 125$ can be read "five **cubed** equals one hundred twenty-five."

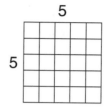

5

The 5 columns and 5 rows form a square.

The depth is also 5. Now we have a cube.

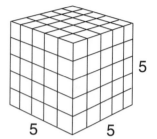

5

5 5

Solving Powers

Examples:

$5^3 = 5 \cdot 5 \cdot 5 = \mathbf{125}$
$25 \cdot 5$

$3^4 = 3 \cdot 3 \cdot 3 \cdot 3 = \mathbf{81}$
$2^5 = 2 \cdot 2 \cdot 2 \cdot 2 \cdot 2 = \mathbf{32}$

  A. Find each value. You may use a calculator. The first one is started for you.

1. $7^2 = 7 \cdot 7 =$ _____ $3^3 =$ _____ $2^3 =$ _____

2. $11^2 =$ _____ $1^5 =$ _____ $12^2 =$ _____

3. $5^3 =$ _____ $8^2 =$ _____ $9^3 =$ _____

4. $4^2 =$ _____ $6^3 =$ _____ $10^3 =$ _____

When a whole number is squared, the result is called a **perfect square.** In the problem $5^2 = 25$, the number 25 is a perfect square.

  B. Complete this table of common perfect squares. These will be useful to know throughout your work in algebra and geometry.

$2^2 =$	$5^2 =$	$8^2 =$	$11^2 =$	$14^2 =$
$3^2 =$	$6^2 =$	$9^2 =$	$12^2 =$	$15^2 =$
$4^2 =$	$7^2 =$	$10^2 =$	$13^2 =$	$20^2 =$

Square Roots

The opposite of raising a number to the second power is finding the **square root.**
The symbol for this operation is called a **radical sign** $\sqrt{}$.

Example: $\sqrt{81} = ?$
Ask yourself, "What number multiplied by itself equals 81?"

The square root of 81 is 9. $\sqrt{81} = \mathbf{9}$

Check:
To check a square root, multiply the answer by itself.

$9 \cdot 9 = 81$, so $\sqrt{81} = 9$

C. **Find the value of each square root. Ask yourself, "What number times itself gives this number?"**

5. $\sqrt{100} =$ _____ $\sqrt{121} =$ _____ $\sqrt{49} =$ _____ $\sqrt{64} =$ _____

6. $\sqrt{144} =$ _____ $\sqrt{169} =$ _____ $\sqrt{400} =$ _____ $\sqrt{81} =$ _____

Many numbers do not have a whole number as a square root. To find the square root of a large number, you can use a calculator. Just enter the number and then press the square root key $\boxed{\sqrt{}}$. If you use a calculator to find the square root of a number that is not a perfect square, the answer will have a decimal part. If you don't have a calculator, you can estimate the square root of a number by comparing it to the square roots you do know.

Example: Find the square root of 110.

You know that 10^2 is 100 and 11^2 is 121. Therefore, you can conclude that $\sqrt{110}$ is between the values of 10 and 11.

D. **Use your knowledge of perfect squares to find the approximate value of each square root.**

7. $\sqrt{50}$ is between

(1) 5 and 6
(2) 6 and 7
(3) 7 and 8

8. $\sqrt{125}$ is between

(1) 10 and 11
(2) 11 and 12
(3) 12 and 13

9. $\sqrt{10}$ is between

(1) 3 and 4
(2) 4 and 5
(3) 5 and 6

10. **Explain** Describe how your knowledge of perfect squares could help you estimate the square root of 90.

Writing Equations

Mathematics is a language for expressing relationships between numbers. The example below shows how a sentence can be translated from words into mathematical symbols.

Example: A number increased by seven is twelve.
$$x \quad + \quad 7 \quad = \quad 12$$

The number sentence $x + 7 = 12$ is an equation. An **equation** is a mathematical statement that two expressions are equal. The $=$ sign separates the two sides of the equation.

$$x + 7 \qquad = \qquad 12$$

left side   equal sign   right side

In this equation, the letter x represents an unknown number. Letters used in equations are called **variables.** Any letter can be used as a variable.

Study the next examples carefully. Notice how the word *is* corresponds to the $=$ sign.

Verbal Expressions	Algebra
The product of five and a number <u>is</u> thirty.	$5c = 30$
Ten less than a number <u>is</u> three.	$y - 10 = 3$
A number divided by eight <u>is</u> seven.	$\frac{a}{8} = 7$
Fifteen increased by a number <u>is</u> twenty.	$15 + w = 20$
Ten decreased by a number <u>is</u> four.	$10 - n = 4$
Three times a number decreased by five <u>is</u> thirteen.	$3m - 5 = 13$

A. Choose the correct equation for each problem.

1. Twelve less than a number is twenty-five.

(1) $12 - x = 25$ **(2)** $x - 12 = 25$ **(3)** $25 - x = 12$

2. Sixty divided by a number is fifteen.

(1) $\frac{60}{n} = 15$ **(2)** $\frac{n}{15} = 60$ **(3)** $\frac{n}{60} = 15$

3. A number increased by nine is twenty.

(1) $9a = 20$ **(2)** $20 + a = 9$ **(3)** $a + 9 = 20$

4. Two less than three times a number is ten.

    (1) $3w - 2 = 10$ **(2)** $2 - 3w = 10$ **(3)** $3 - 2w = 10$

5. The product of five and a number increased by two is twelve.

    (1) $2b + 5 = 12$ **(2)** $5b + 2 = 12$ **(3)** $b + 2 + 5 = 12$

6. Ten divided by a number is three decreased by that number.

    (1) $\frac{10}{z} = 3 - z$ **(2)** $\frac{z}{10} = z - 3$ **(3)** $\frac{10}{2} = 3 - z$

B. Write a verbal expression for each equation.

7. $5 + x = 8$ _____

8. $\frac{12}{a} = 6$ _____

9. $4y - 5 = 7$ _____

C. Choose an equation for each situation described below.

10. Warren has $320 in his checking account. After making a deposit (d), he has $400 in the account.

    (1) $\$320d = \$400$

    (2) $\$320 + d = \$400$

    (3) $\$320 - d = \$400$

11. Four times the measure of the side (s) of a square equals a perimeter of 12 inches.

    (1) $\frac{4}{s} = 12$

    (2) $4 + s = 12$

    (3) $4s = 12$

12. The sale price of a coat, which is $85, is $35 less than the original price (p).

    (1) $p + \$35 = \$85$

    (2) $p - \$35 = \$85$

    (3) $\$35 - p = \$85$

13. Socks sell for $1.15 per pair. The total cost before tax for a number of pairs (n) of socks is $5.75.

    (1) $\$1.15n = \$5.75$

    (2) $\$1.15 + n = \$5.75$

    (3) $\frac{\$1.15}{n} = \$5.75$

SOCKS
$1.15 per pair

14. For her job, LeeAnn drove the same distance (d) each day for 5 days for a total distance of 1,650 miles.

    (1) $5d = 1{,}650$

    (2) $5 + d = 1{,}650$

    (3) $\frac{5}{d} = 1{,}650$

15. Explain Two students, Sam and Jin, are given this situation: The $120 charge for computer repairs is $30 more than twice the technician's hourly rate (r). Sam writes the equation $\$120 = \$30 + 2r$. Jin writes the equation $2r + \$30 = \$120$. Explain why both students are correct.

Order of Operations

Look at the expression $3 + 4 \times 5$. To **evaluate** the expression means to solve the expression. You must decide whether to first add $3 + 4$ or to first multiply 4×5. If you add first, you get $7 \times 5 = 35$. If you multiply first, you get $3 + 20 = 23$. Both answers cannot be right.

In order to make sure that everyone solving an expression gets the same answer, mathematicians developed an order to follow.

Using the Order of Operations

To evaluate expressions, perform the operations in the following order. Within each level, always work from *left to right*.

1. Grouping symbols: parentheses, brackets, and fraction bars
2. Powers and roots
3. Multiplication and division
4. Addition and subtraction

$$(7 - 3)^2 + 3 \cdot 4$$
$$(4)^2 + 3 \cdot 4$$
$$16 + 3 \cdot 4$$
$$16 + 12$$
$$\mathbf{28}$$

Notice how the numbers are brought down in each line of work in this example. The 4 is under $7 - 3$, and 16 is under $(4)^2$. You should get in the habit now of lining up evaluation problems carefully.

Following the Order of Operations

Example: Evaluate the expression $\frac{10 - 4}{2}$.

Step 1
Do the operation above the fraction bar.

Hint: When working with fraction bars, do all of the work *above* and *below* the fraction bar before dividing the top by the bottom.

$$\frac{10 - 4}{2}$$
$$\frac{6}{2}$$
$$3$$

Step 2
Divide.

The solution is **3.**

Example: Art has a test Friday afternoon. He plans to study 4 hours on Monday and 2 hours per day on Tuesday through Friday. Write and evaluate an expression to find how many hours Art plans to study for the test.

Step 1
Write the expression.

$$4 + 2(4)$$

Step 2
Since multiplication comes before addition in the order of operations, multiply first.

$$4 + 8$$
$$12$$

Step 3
Add.

The solution is **12 hours.**

A. Evaluate each expression.

1. $4 \cdot 5 + 3 \cdot 2$ $\frac{10 + 15}{10 - 5}$ $2 \cdot 4^2 + 12$

2. $9 - 7 + 6 - 5$ $3(8 - 4)$ $5(2 + 7)^2$

3. $\sqrt{81} - 5$ $5 \cdot 3^2$ $6 \cdot 2 - (5 + 6)$

4. $\frac{24}{6} - \frac{36}{12}$ $\frac{40 - 4}{9}$ $\frac{60}{20 - 8}$

5. $(3 + 8)^2$ $3(8 + 4) - (15 + 9)$ $9(12 - 9)^2$

B. For each problem, choose the expression that shows the result after the *first operation* is completed.

6. $7 \cdot 6 - 5 \cdot 4$
(*Hint:* Multiplication precedes subtraction.)
  (1) $7 - 1 - 4$
  (2) $42 - 20$
  (3) $7 + 1 - 4$

8. $2(9 - 6)$
  (1) $18 - 6$
  (2) $2(15)$
  (3) $2(3)$

10. $(8 + 5)(8 - 5)$
  (1) $13(3)$
  (2) $8 + 40 - 5$
  (3) $13 - 13$

7. $\frac{20 + 40}{10} - 5$
  (1) $\frac{60}{10} - 5$
  (2) $\frac{60}{5}$
  (3) $20 + 4 - 5$

9. $\frac{21 + 18}{3}$
  (1) $(21)(18)$
  (2) $\frac{39}{3}$
  (3) $\frac{21 - 18}{3}$

11. $15 + \sqrt{36}$
  (1) 51
  (2) $15 + 36$
  (3) $15 + 6$

C. Read the situation below and write an expression that could be used to solve it.

12. Three families are going to a play. They decide to split the cost of the tickets evenly. The families need to buy 7 tickets for children at $9 each and 6 tickets for adults at $12 each. How much does each family owe?

13. Analyze Is the order of operations important in solving problem 12? Try performing the operations out of order to see if the answer is affected.

The Distributive Property

At the $3 Discount Store, all items cost $3. Maia buys 6 T-shirts and 4 pairs of jeans for her sons. What is the cost of the clothing before sales tax?

If you think carefully about this problem, you will see that there are two ways to solve it.

Method 1. First find the total number of items. Then multiply the number of items by $3, the cost per item.

$6 + 4 = 10$ items
$3(10) = \$30$

Method 2. First find the cost of the T-shirts and jeans separately. Then add to find the total cost.

$3(6$ T-shirts$) = \$18$
$3(4$ jeans$) = \$12$
$\$18 + \$12 = \$30$

Each method gives an answer of $30.

Now think about how you would write the numerical expression for each method.

Method 1. $3(6 + 4)$ The parentheses are used as both grouping and multiplication symbols. First add within the parentheses, then multiply.

Method 2. $3(6) + 3(4)$ These parentheses mean multiplication.

Since the value of both expressions is $30, we can say the two expressions are equal.

$$3(6 + 4) = 3(6) + 3(4)$$

This example demonstrates the **distributive property.** We say that *multiplication is distributive over addition and subtraction.* In this example, 3 is distributed over the sum of 6 and 4.

We can also show the distributive property using variables.

Addition
$$a(x + y) = ax + ay$$
$$2(4 + 3) = 2(4) + 2(3)$$
$$2(7) = 8 + 6$$
$$14 = 14$$

In these examples, $a = 2$, $x = 4$, and $y = 3$.

Notice how the left side of each equation shows Method 1 and the right side shows Method 2. Both methods are correct and equal.

Subtraction
$$a(x - y) = ax - ay$$
$$2(4 - 3) = 2(4) - 2(3)$$
$$2(1) = 8 - 6$$
$$2 = 2$$

A. Choose the correct application of the distributive property.

1. $3(8 - 5)$
  (1) $3 \cdot 8 - 5$
  (2) $3 \cdot 8 - 3 \cdot 5$
  (3) $3 + 8 - 3 \cdot 5$

2. $c \cdot 15 - c \cdot 10$
  (1) $c(15 - 10)$
  (2) $(c + 15) - (c + 10)$
  (3) $c(15 + 10)$

3. $9 \cdot 5 + 9 \cdot 6$
  (1) $9(5 + 6)$
  (2) $(9 + 5)(9 + 6)$
  (3) $9(5) + 6$

4. $6(x + 12)$
  (1) $6 + 12x$
  (2) $6x + 12$
  (3) $6x + 6 \cdot 12$

B. Do not solve the problems below. Instead, for each problem, write two expressions that could be used to solve it.

5. For the office, Max bought 2 ink-jet printers and 2 ink cartridges. The printers cost $550 each and the cartridges cost $28 each. What was the total cost of the printers and the cartridges?

6. Carol used to earn $11 per hour. Now she earns $16 per hour. How much more does she earn now for a 40-hour workweek?

7. Film is sold in packs of 3 rolls. Cathy bought 3 packs of Super Gold film and 6 packs of True Blue film. How many more rolls of True Blue film did she buy?

8. Laurie bought 4 bottles of correction fluid, each containing 0.7 fluid ounce. She also bought 4 refill bottles, each containing 4.5 fluid ounces. How many fluid ounces of correction fluid did she buy?

For another look at order of operations, turn to page 270.

Making Connections: Perimeter of a Rectangle

Perimeter is the distance around something. The formula for finding the perimeter of a rectangle is $P = 2l + 2w$, where P = perimeter, l = length, and w = width.

11 cm

8.5 cm

1. Use the formula to find the perimeter of the rectangle.

2. Using the distributive property, find another way to write the formula for finding the perimeter of a rectangle.

Addition and Subtraction Equations

Think about this equation: $5 + x = 9$. Since you know that the sum of 5 and 4 is 9, you know the value of x must be 4. It is not difficult to see the solution to an equation with small whole numbers and one operation. But to find the solution to a complicated equation, you need a method.

An equation is solved when you find the value for the variable.

A solution can take either of these forms.

$$x \quad = \quad 10 \qquad \text{or} \qquad 10 \quad = \quad x$$
unknown   value   value   unknown

To solve an equation, you want to get the variable alone on one side of the equal sign. To do this, use inverse operations.

For example, if you have $100 in savings and you spend $20 (subtract), how do you get back to your original balance? You must deposit (add) $20.

Addition and subtraction are inverse operations. If you subtract 20, you can undo the operation by adding 20.

The important thing to remember is that, whatever you do to one side of an equation, you need to do the same to the other side to keep things equal. Think of an equation as a balance scale that must be balanced.

Using Inverse Operations

Example 1: Solve for y in the equation $y + 16 = 64$.

Think: You want to get y alone. The operation in the equation is addition. The inverse of addition is subtraction.

Do: Subtract 16 from both sides.
The solution is $y = $ **48.**

$$\begin{aligned} y + 16 &= 64 \\ -16 \quad &-16 \\ \hline y &= 48 \end{aligned}$$

Check: Substitute 48 for y in the equation.

$$48 + 16 = 64$$
$$64 = 64$$

Example 2: Solve for n in the equation $3.25 = 2.4 + n - 9.2$.

Think: You want to get n alone.
First, combine like terms: $2.4 - 9.2 = -6.8$
The operation in the equation is subtraction.
The inverse operation is addition.

Do: Add 6.8 to both sides.
The solution is $n = $ **10.05.**

$$\begin{aligned} 3.25 &= n - 6.8 \\ +6.8 \quad & \quad +6.8 \\ \hline 10.05 &= n \end{aligned}$$

Check: Substitute 10.05 for n in the equation.

$$3.25 = 10.05 - 6.8$$
$$3.25 = 3.25$$

A. Solve and check each equation.

1. $a - 15 = 78$ $5.3 = x + 0.6$ $532 = 189 + n$

2. $b + 3\frac{1}{2} = 9$ $10 = w - 8.2$ $d - 11 = 33$

3. $x + 6.4 = 100$ $1,050 + g = 8,000$ $9\frac{1}{2} = y - 3\frac{1}{2}$

B. Simplify each equation by combining numerical terms. Then solve each equation. The first one is started for you.

4. $c + 28 - 7 = 84$ $h + 1.5 + 8.75 = 12.5$
$c + 21 = 84$

5. $95 - 25 + 15 + x = 105$ $5.375 = e - 0.9 + 2.05$

C. For each problem below, first write an equation. Then solve for the unknown.

6. According to the map below, Leon has driven 265 miles of a 770-mile trip. Let m equal the remaining miles he has to drive. Solve for m.

8. The perimeter (distance around) the triangle shown below is 6.5 centimeters. What is the measure of side a?

7. Nita earns $420 per week. The total of her and her husband Craig's weekly salaries is $795. Let c equal Craig's weekly earnings. How much does Craig earn per week?

9. **Explain** The difference in the high and low temperatures for Provo on January 9 was 24°. The high temperature was 18°F. Let l represent the low temperature. Explain how you know that l will be a negative number. Solve for l.

Multiplication and Division Equations

The tag on a team jacket shows both the original price and the sale price. The original price is 3 times greater than the sale price. To find the original price from the sale price, you could multiply the sale price by 3. To find the sale price from the original price, you could divide the original price by 3.

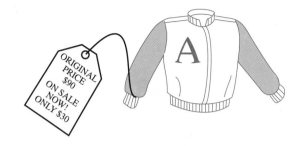

Using Inverse Operations

Example 1: Solve for z in the equation $4z = 32.8$.

Think: You want to get z alone. The operation in the equation is multiplication. The inverse of multiplication is division.

Do: Divide both sides of the equation by 4. (Remember, $4z \div 4 = 1z$ and $1z = z$.)
The solution is $z = \textbf{8.2.}$

$$\frac{4z}{4} = \frac{32.8}{4}$$
$$z = 8.2$$

Check: Substitute 8.2 for z in the equation. $4(8.2) = 32.8$, and $32.8 = 32.8$

Example 2: Solve for t in the equation $15 = \frac{t}{9}$.

Think: You want to get t alone. The operation in the equation is division. The inverse operation is multiplication.

Do: Multiply both sides by 9.
The solution is $\textbf{135} = \textbf{t.}$

$$15(9) = \left(\frac{t}{9}\right)(9)$$
$$135 = t$$

Check: Substitute 135 for t in the equation. $15 = \frac{135}{9}$, and $15 = 15$

A. Solve and check each equation.

1. $\frac{w}{6} = 9$ $12n = 600$ $\frac{z}{16} = 128$

2. $1.5x = 45$ $\frac{p}{2} = 56$ $25r = 40$

3. $50c = 3,000$ $\frac{y}{0.5} = 150$ $\frac{h}{32} = 3$

Some equations require more than one step to find a solution. When you need to perform more than one operation to find a solution, reverse the order of operations. Do any addition or subtraction steps first. Then multiply or divide.

Using Two Inverse Operations

Example: Solve the equation $6x + 5 = x + 25$.

Step 1. Subtract x from each side.
Notice that $x = 1x$.
Subtract 5 from each side.

$6x - x + 5 = x - x + 25$
$5x + 5 = 25$
$5x + 5 - 5 = 25 - 5$
$5x = 20$

Step 2. Divide both sides by 5.

$\frac{5x}{5} = \frac{20}{5}$
$x = 4$

Check: Substitute 4 for x in the equation.

$6(4) + 5 = 4 + 25$
$24 + 5 = 29$
$29 = 29$

B. Solve and check each equation.

4. $7x + 3 = 17$ $\qquad$ $\frac{y}{6} - 3 = 1$ $\qquad$ $10x - 6 = 24$

5. $3b + 7 = b - 9$ $\qquad$ $4n - 15 = 7n - 51$ $\qquad$ $\frac{z}{3} - 4 = 2$

C. Write an equation and then solve for the variable.

6. Stuart sells newspaper advertising. He earns $200 per week plus a $15 commission for every ad he sells. Last week he earned $470. Let a represent the number of ads. How many ads did Stuart sell?

7. Last week, Leslie worked 4 hours more than twice the number of hours John worked. Let h be the number of hours John worked. If Leslie worked 38 hours, how many hours did John work last week?

8. Last season, the Hornets won 3 times as many games as they lost. If they played 60 games, how many did they lose? (*Hint:* Add the games won $3x$ to the games lost x to get the total 60.)

9. The perimeter of the rectangle below is 96 inches. As shown in the diagram, the length of the rectangle is 36 inches. Find the width (w) of the rectangle.

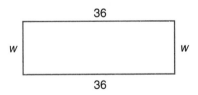

10. **Explain** The formula for finding the perimeter of a rectangle is $P = 2l + 2w$, where P = perimeter, l = length, and w = width. Explain how to use the formula in solving problem 9.

Working with Formulas

Formulas are useful tools that help us remember how to solve certain kinds of problems. A **formula** is an equation that shows a constant relationship between variables.

For example, the formula for finding the perimeter (P) of a square is
  $P = 4s$, where s = side

The formula reminds us that the perimeter of a square is equal to four times the length of one side. Using the formula, we can find that the perimeter of the square to the right is 20.

Formulas are very useful. Unfortunately, they are not always written in the form you need to solve a problem. You can use what you know about equations to write a useful equivalent formula.

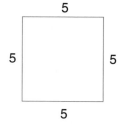

$P = 4s$
$P = 4(5)$
$P = 20$

Working with Equivalent Formulas

Example: You know the perimeter (P) of a square is 20. You need to find the length of one side (s). Rewrite the formula $P = 4s$ so that you can solve for s.

Step 1. Write an equivalent formula. $\frac{P}{4} = \frac{4s}{4}$

Think: You want to get s alone. The operation in the equation is multiplication. The inverse operation is division.

Do: Divide both sides of the equation by 4. $\frac{P}{4} = s$

Step 2. Use the new formula. Substitute 20 for P. $\frac{20}{4} = 5$
The solution is $s = $ **5.**

A. Write equivalent formulas to solve the following.

1. **a.** The formula for finding the area (A) of a rectangle is
    $A = lw$, where l = length and w = width. Write an equivalent
    formula to solve for w. (*Hint:* Use the inverse of multiplication.)

   b. Find the width of a rectangle with an area of 180 square inches
    and a length of 15 inches.

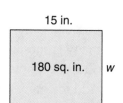

174

2. a. The formula for finding the perimeter (P) of a triangle is
$P = a + b + c$, where a, b, and c are the sides of the
triangle. Write an equivalent formula to solve for c.

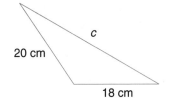

b. The perimeter of the triangle to the right is 72 centimeters.
Find the length of side c.

3. a. The formula for finding the mean (x) or average of two numbers is
$x = \frac{a + b}{2}$, where a and b represent the two numbers. Write an
equivalent formula to solve for a. (*Hint:* Multiply both sides first
to undo the division operation.)

b. The average of two numbers is 65. One of the numbers is 50.
Find the other number.

Making Connections: The Distance Formula

You worked with the distance formula in Unit 3. Remember, the formula $d = rt$ shows
the relationship between distance (d), rate or speed (r), and time (t).

Example: An airliner travels an average speed of 600 miles per hour
for 3 hours. How many miles does the airliner travel? To solve for
distance (the number of miles traveled), multiply the rate (600 MPH)
by the time (3 hours).

$$d = rt$$
$$d = 600(3)$$
$$d = \textbf{1,800 miles}$$

You can also write equivalent equations to solve for either rate or time.

Example: Elaine needs to make a trip to Carson, a distance of 175 miles. She expects
to average 50 miles per hour. At that rate, how long would the trip take?

Step 1. Write an equivalent formula to solve for t. $\qquad \frac{d}{r} = t$
Divide both sides by r to get t alone. $\qquad \qquad \frac{175}{50} = t$

Step 2. Use the new formula. $\qquad \qquad \qquad 3\frac{1}{2}$ or **3.5 hours** $= t$

Use a form of the distance formula to solve these problems.

1. On a mountain road, Jeff drives 20 miles per hour. How far can he travel in $2\frac{1}{2}$ hours?

2. A model airplane traveled 60 feet in 12 seconds. What was the speed of
   the plane? (*Hint:* Your answer will be in feet per second.)

Substituting to Solve Equations

So far you have used substitution to evaluate expressions and to check answers.
You can also use substitution to solve problems.

Substituting a Value in an Equation

Example: What is the value of x if $4x + 3y = 59$ and $y = 5$?

Step 1. Substitute 5 for y.	$4x + 3(5) = 59$
Step 2. Subtract 15 from both sides.	$4x + 15 = 59$
Step 3. Divide both sides by 4.	$4x = 44$
The value of x is **11**.	$x = 11$

A. Solve each problem using substitution.

1. What is the value of x if $3x - y = 12$ and $y = 6$? (*Hint:* Substitute the known value for y in the first equation.)

2. Find the value of n when $2m + 2n = 10$ and $m = 3$.

3. What is the value of a if $5a + 2c = -12$ and $c = -1$?

4. Find the value of t when $4(s + 4t) = 100$ and $s = 5$.

Sometimes you can solve for a variable and use that solution to solve another equation.

Solving for One Value to Find Another

Example: Find the values of a and b if $3a - 16 = 20$ and $b = \frac{a}{4}$.

Step 1. First solve for a (get a alone).	$3a - 16 = 20$
Step 2. Add 16 to both sides.	$3a = 36$
Step 3. Divide both sides by 3. Now use the value of a to solve for b.	$a = 12$ $\quad b = \frac{a}{4}$
Step 4. Substitute 12 for a.	$b = \frac{12}{4}$
Step 5. Divide.	$b = 3$

B. Find the values of both variables.

5. Solve for r and s if $3r = r - 8$ and $s = -2r$.

6. Find w and v if $7w + 10 = 80$ and $2v = 3w$.

7. Solve for c and d if $10c - 6 = 2(c + 1)$ and $2d = 8c$.

8. Solve for x and y if $9(x + 4) = 4x - 39$ and $45 - 2x = y$.

In Part B, the first solution was a number that was then substituted in an expression. In Part C, substitute an expression instead of a number.

Substituting an Expression for a Variable

Example: Find the values of x and y if $4x + 5y = 7$ and $y = 3x + 9$.

Step 1. Substitute $3x + 9$ for y in the first expression.	$4x + 5(3x + 9) = 7$
Step 2. Simplify with the distributive property.	$4x + 15x + 45 = 7$
Step 3. Combine like terms.	$19x + 45 = 7$
Step 4. Subtract 45 from both sides.	$19x = -38$
Step 5. Divide both sides by 19.	$x = -2$
Step 6. To find y, substitute -2 for x in the second expression.	$y = 3(-2) + 9$ $y = -6 + 9 = 3$

The solutions are $x = -2$ and $y = 3$.

C. Find the values of both variables.

9. Find the values for a and b if $2a + 7b = 78$ and $a = b - 6$.

10. Solve for d and e if $4e + 2d = -6$ and $e = d + 9$.

D. Solve each problem using your knowledge of substitution.

11. Mary's hourly wage m is twice Tom's hourly wage t ($m = 2t$). Last week Mary earned $436 for 40 hours work plus $16 in overtime ($40m + \$16 = \$436$). How much does Tom earn per hour?

13. The perimeter of the rectangle shown below is 260. Find the length and width of the rectangle if $y = 20$.

$x + y$

$5y$

12. The distance v from Riverside to Visalia is 3 times the distance b from Riverside to Barstow ($v = 3b$). The difference in mileage between the two trips is 160 miles ($v - b = 160$). How far is it from Riverside to Barstow? ($b = ?$)

14. **Explore** Do you need to know the perimeter of the rectangle to solve problem 13? Explain your reasoning.

Writing and Solving Inequalities

You know that < means "is less than" and > means "is greater than." Remember, the symbol points to the smaller number. These symbols and others in the list to the right are commonly used to express inequalities.

Symbol	Meaning	Example
<	is less than	$-8 < -1$
>	is greater than	$2 > -3$
≤	is less than or equal to	$1 \leq 3$
≥	is greater than or equal to	$4 \geq 4$

Solving inequalities is similar to solving equations.

Solving and Graphing Inequalities

Example: Solve the inequality below.

$$2y - 1 < 5$$

Add 1 to both sides.
$$2y - 1 + 1 < 5 + 1$$
$$2y < 6$$

Divide both sides by 2.
$$\frac{2y}{2} < \frac{6}{2}$$

The solution is **y < 3.** $\quad y < 3$

Check: The inequality is true if $y = 2$ because $2 < 3$.

We can graph the solution on a number line. The highlighted portion of the line shows that the solution set includes all numbers less than 3. The mark representing 3 is circled (not shaded) because 3 is not a possible solution.

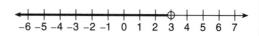

A. **Solve and graph each inequality. The first one is done for you.**

1. $4b - 3 \geq 9$

$4b \geq 12$

$b \geq 3$

The closed circle shows that the solution set includes 3.

$2(x + 5) \leq 18$

Hint: Use the distributive property to remove the parentheses.

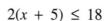

$12 > 6m + 2 - m$

2. $2n + 8 > 4$

$6 \leq 3(y - 2)$

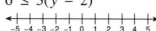

$7 < 3a - 1 + a$

B. For problems 3–5, choose the values for which each inequality is true. Each problem may have more than one solution.

3. $y - 4 \leq 3$ **(1)** 10   **(2)** 8   **(3)** 7   **(4)** 5   **(5)** 2   **(6)** –1

4. $b + 2 > 1$ **(1)** 3   **(2)** 1   **(3)** 0   **(4)** –1   **(5)** –3   **(6)** –5

5. $2x - 10 \geq -2$ **(1)** –8   **(2)** –4   **(3)** 0   **(4)** 4   **(5)** 8   **(6)** 12

C. Write an inequality for each statement. Let x represent the unknown value. Solve for x. The first one is started for you.

6. Ten more than a number is greater than or equal to 6.

$$x + 10 \geq 6$$

7. Five more than a number is greater than negative 3.

8. A number decreased by 4 is less than 2 times that number decreased by 10.

9. Two times a number, increased by 18, is less than or equal to 8 times that number.

D. Frank is picking up food at a fast food restaurant to take to a potluck picnic. He doesn't want to spend more than $15 before tax. For each problem below, write an inequality and then solve for the unknown. Use the information from the menu. You may use a calculator.

10. For $15, how many Super Burgers with cheese could Frank buy?

11. Frank decides to buy a small order of fries and a large drink with each burger. If he buys Junior Burgers without cheese, how many meals can he buy?

12. Frank changes his mind again. If he buys himself a Super Burger with cheese, large fries, and a large drink, how many Junior Meals can he buy in addition without going over budget?

13. **Compare** Frank can buy a gallon of soda for $5. Frank estimates that 8 large drinks = 1 gallon. Write > or < in the space below to make a true statement.

Cost of 1 gallon of soda _____ Cost of 8 large drinks

 Zestos

Super Burger.........................	$1.99
with cheese...................	$2.19
Junior Burger........................	$1.49
with cheese...................	$1.69
Large Fries	$.99
Small Fries	$.59
Milkshake	$1.19
Large Drink	$1.09
Small Drink............................	$.79
Junior Meal	$2.99

includes burger, fries, drink, and toy

Translating Words to Equations

It is sometimes helpful to break down information into smaller parts.
You can sometimes use charts and diagrams to organize these parts.

Using a Chart to Organize Information

Example: Laura, Robin, and Marilyn start a dog obedience school together. Because they will not be investing equal amounts of time and money, they will not split the profits evenly. They agree that Robin will receive twice what Laura earns, and Marilyn will receive twice Robin's earnings. How much is each person's share of $3,500?

Create a chart showing what you know. You know the least about Laura's share. Let x represent her share. Now fill out the rest of the chart using information from the problem.

Laura's Share	Robin's Share (2 times Laura's share)	Marilyn's Share (2 times Robin's share)
x	$2x$	$2(2x)$

Step 1
Write an equation using the chart.
Remember, the shares total $3,500.

$$x + 2x + 2(2x) = \$3,500$$
$$x + 2x + 4x = \$3,500$$

Step 2
Then solve for x.

$$7x = \$3,500$$
$$x = \$500$$

Step 3
Go back to the chart. Laura's share is **$500,** the value of x.
Substitute $500 for x to find Robin's share: 2($500) = $1,000. Robin's share is **$1,000.**
You know that Marilyn's share is 2 times Robin's share: 2($1,000) = $2,000. Marilyn's share is **$2,000.**

A. Create a chart for each problem and solve.

1. David got 2 tickets. The speeding ticket was $6 more than 3 times the cost of the parking ticket. Together he paid $78 in fines. How much was the parking ticket? (*Hint:* Let x = parking ticket.)

Parking	Speeding

2. Al, Art, and Anna work in a store. Last week, Art worked 6 hours less than twice Al's hours. Anna worked 10 hours more than Art. Altogether, they worked 103 hours. How many hours did Art work last week?

Al	Art	Anna

Age problems seem tricky because they deal with the present and the future. Use a chart to organize the information.

B. Complete each chart. Then solve for the unknown. Make sure you answer the question asked. The first one is started for you.

3. Cathy is 5 times as old as her daughter Maggie. In 10 years, the sum of their ages will be 68. How old is Maggie now?

	Cathy	Maggie
Age Now	$5x$	x
In 10 Years	$5x + 10$	$x + 10$

$$5x + 10 + x + 10 = 68$$

4. Chris is 4 years older than his sister Ann. Two years from now, Chris will be twice as old as Ann. How old is Ann now?

	Chris	Ann
Age Now		
In 2 Years		

With money problems, write algebraic expressions for the number of each coin or bill. The value of a number of quarters, based on pennies, can be expressed as $25x$. The value of a number of dimes could be $10x$.

C. Complete the chart. Then solve for the unknown. Make sure you answer the question asked in the problem.

5. Yuki has a total of $1.60 in dimes and nickels in her pocket. She has twice as many nickels as dimes. How many does she have of each?

	Dimes	Nickels
Number of Coins	x	$2x$
Value	$10x$	

7. Phil put $5.50 in parking meters today. He used only quarters and dimes. If he used 4 times as many quarters as he did dimes, how many quarters did he spend today?

	Dimes	Quarters
Number of Coins		
Value		

6. Risa has an equal number of $1, $5, and $10 bills in her wallet. If she has $80 in her wallet, how many of each bill does she have?

	$1 bills	$5 bills	$10 bills
Value			

8. Write Make up a problem about the number of $20, $10, and $5 bills in a cash register drawer. Design a chart that could be helpful in solving the problem.

Mixed Review

A. Write an expression for each problem. Do not solve the problem.

1. SooJung paid $8 per class for 6 dance classes.

2. Three roommates shared the cost of the $78 utility bill equally.

3. The value of the baseball card now is $50 greater than its 1969 value of $5.

4. The $550 cost of the printer was decreased by the $100 rebate.

B. Evaluate each problem.

5. 2^4

6. 3^3

7. 13^2

C. Choose the correct solution to each problem.

8. $\sqrt{400}$
   (1) 20   **(2)** 40   **(3)** 200

9. $\sqrt{260}$
   (1) between 14 and 15   **(2)** between 15 and 16   **(3)** between 16 and 17

10. $\sqrt{30}$
    (1) between 5 and 6   **(2)** between 6 and 7   **(3)** between 7 and 8

D. Fill in each blank with a symbol (<, >, or =) that makes the statement true.

11. –5 _____ –1

12. +6 _____ 6

13. 2 _____ –2

E. Solve each of the following.

14. 18 + (–12) =

16. 8 – (–10) =

18. 2 + (–3) + 1 =

15. –4 + (–12) =

17. 8 + (–20) – 13 =

19. 100 – (+50) – (–75) =

182

F. Write an algebraic expression for each situation. Do not solve.

20. Ana had n dollars. She spent $5.85. How much money does she have left?

21. Timothy is 10 inches less than twice the height (h) of his son. Write an expression describing Timothy's height.

22. How many batteries can be bought for x amount of money, if one battery costs y?

23. The winner received 3,500 more than 3 times the number of votes received by the loser. Let v = the loser's votes. Write an expression describing the winner's votes.

G. Solve each equation or inequality.

24. $4x - 5 = 7$

25. $12 - 4y = -4$

26. $3x - 1 > 6x - 7$

27. $4a - 9 = 8a + 3$

28. $6n + 15 \geq 8n - 1$

29. $3z - 8 < 10z - 2 - 8z$

H. Use substitution to solve the following.

30. Find the value of y if $5x^2 - y = 20$ and $x = 5$.

31. Find the value of a and b if $4a - 12 = 0$ and $b = 2a + 4$.

32. Solve for c and d if $-4c + 2 = 2(c - 8)$ and $\frac{1}{3}d = 3c$.

33. Solve for m and n and $5m + 3n = 26$ and $m = n + 2$.

I. For each problem below, first write an equation. Then solve for the unknown. You may use a calculator.

34. Kent has a board that is 84 inches long. He needs to cut it into 2 pieces so that one piece is 10 inches longer than the other piece. Find the length of each piece.

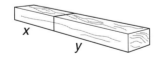

35. The Epp family's gas bill for November was $8 more than twice the bill for July. If the total of the two bills was $56, how much was the gas bill in November?

36. There are 85 students in three classes. Class A has 12 more students than Class B. Class C has 23 fewer than twice the number in Class B. How many students are in each class?

Class A	Class B	Class C

Points, Lines, and Angles

Scientists identify materials by their *properties*. For example, one property of rubber is called elasticity (the ability of a material to return to its original shape). In geometry, we also speak about properties.

A **point** can be described by its position in space.

A **ray** is a straight path of points that starts at one point and continues infinitely in one direction.

A **line segment** has definite length. In the illustration, the line segment lies between its **endpoints** A and B.

A **line** is a straight path of points that continues in two directions. A line has infinite length. In other words, it goes on and on. Arrows are used to show that the line continues in both directions. Often the word *line* is also used when talking about a ray or a line segment.

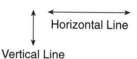

Some lines belong in special categories. **Vertical** lines run straight up and down. **Horizontal** lines run left and right.

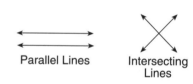

Two or more lines that run in the same direction are called **parallel** lines. No matter how far parallel lines are extended, they never cross, or **intersect.**

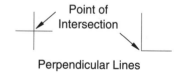

Lines that intersect to form square corners (also called right angles) are called **perpendicular** lines.

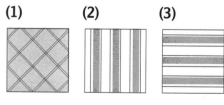

A. **Use what you know about geometric properties to solve the following problems. Choose the correct illustration for each problem.**

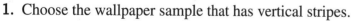

1. Choose the wallpaper sample that has vertical stripes.

2. Which wallpaper sample has lines that intersect?

3. The bracing on the platforms is shaded. Which platform has braces that are perpendicular?

4. Which platform has parallel braces?

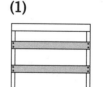
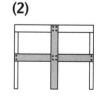

Angles

A **vertex** is a point where two rays or line segments intersect. An angle is formed by two rays. The rays that form an angle are called the sides of the angle.

The size of an angle depends on the amount of **rotation** of the sides. Imagine a clock with both hands pointing at the 12. As one hand moves clockwise, the measure of the angle created by the hands increases. Angles are measured in degrees (°). One full rotation (forming a complete circle) is 360°.

The angle increases as the hand moves.

Below are the properties of four angles that you will need to know.

Tip
A right angle is indicated by a small box drawn inside the angle.

Name of Angle	Properties	Examples
right angle	exactly 90°	90° 90°
acute angle	less than 90°	30° 60° 45°
straight angle	exactly 180°	180° 180°
obtuse angle	between 90° and 180°	120° 140°

B. Answer each question.

5. Match the illustrations with the descriptions.

____ an acute angle
____ a right angle
____ an obtuse angle
____ a straight angle

a.

b.

c.

d.

6. Which of the three angles is smallest? Why? (*Hint:* The *size* of an angle is based only on the angle, not the length of the sides.)

(1) (2) (3)

7. **List** Look around your present surroundings. Find and list an example of each of the following: perpendicular lines, parallel lines, a right angle, an acute angle, an obtuse angle, and a straight angle.

Protractors

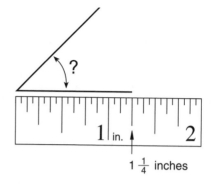

You have already seen how a ruler can be used to measure the length of a line segment. But how can you measure the angle created by two line segments?

$1\frac{1}{4}$ inches

A **protractor** is a tool for measuring angles. A protractor looks like a half circle with a fanlike scale of numbers. One scale runs from left to right; the other from right to left.

To measure an angle, line up one ray of the angle with the baseline of the protractor. Put the center point of the protractor (sometimes marked with crosshairs) on the vertex of the angle. Read the scale at the point where the other ray crosses it.

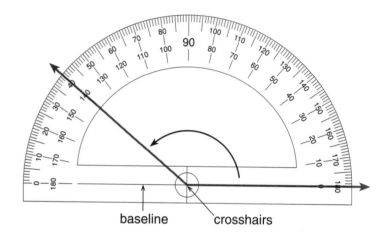

baseline   crosshairs

This illustration shows a protractor and an angle of 140°. The symbol ∠ means angle. Angles can be named by the vertex point, by a letter or number within the opening, or by three points.

The angle at the right could be named ∠A, ∠1, ∠BAC, or ∠CAB. When three points are used to name an angle, the vertex is always the middle letter.

In this illustration, ∠NOP is divided into two smaller angles, ∠x and ∠y. If ∠x = 35° and ∠y = 30°, we can add to find the measure of ∠NOP.

35° + 30° = 65°   ∠NOP = 65°

A. Answer each question.

1. Tell the number of degrees in each of the following angles.

a.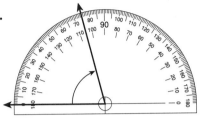

b.

c.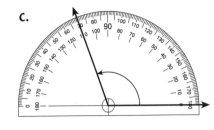

d.

2. Match the angle with the correct number of degrees. Use your judgment or a protractor if you have one.

_____ 180°

_____ 90°

_____ 30°

_____ 135°

a.   b.   c.   d.

B. Solve the following problems without using a protractor.

3. ∠FGH = 160°. Find the measure of ∠JGH.

4. Tell the type of angle and number of degrees for each angle:

Angle	Type	Degrees
∠POS		
∠QOS		
∠POR		

5. ∠AOB = 120° and ∠c is 3 times as great as ∠d. Use algebra to find the measurement of both ∠c and ∠d. (*Hint:* Let a variable equal the measurement of ∠d.)

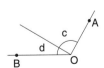

6. **Explore** As you have already seen, the hands on a clock form angles as the hands move around the clock face. During a 24-hour period, how many times do the hands form a straight angle (180°)?

Types of Angles

The right angle pictured here is divided into two angles, $\angle s$ and $\angle t$. If we know the measure of one of the angles, we can find the other by subtracting from 90°, the measure of the right angle.

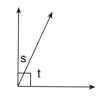

▶ Two angles that add up to 90° are called **complementary angles.**

Example: If $\angle s$ in the figure above measures 25°, what is the measure of $\angle t$?

You know that $\angle s + \angle t = 90°$. Therefore, $90° - \angle s = \angle t$
Subtract 25° from 90°: $90° - 25° = 65°$. $\angle t = \mathbf{65°}$
$\angle t$ is the complement to $\angle s$.

The straight angle pictured here is divided into two angles, $\angle a$ and $\angle b$. If we know the measure of one of the angles, we can find the other by subtracting from 180°, the measure of a straight angle.

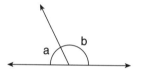

▶ Two angles that add up to 180° are called **supplementary angles.**

Example: If $\angle a$ measures 63°, what is the measure of $\angle b$?

You know that $\angle a + \angle b = 180°$. Therefore, $180° - \angle a = \angle b$
Subtract 63° from 180°: $180° - 63° = 117°$. $\angle \mathbf{b} = \angle \mathbf{117°}$
$\angle b$ is the supplement to $\angle a$.

When two straight lines intersect, they form two pairs of supplementary angles. The angles that share a side are called **adjacent angles.** In this figure, $\angle c$ and $\angle d$ are adjacent. Because they combine to form a straight angle, they are also supplementary.

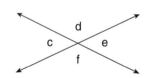

Angles that do *not* share a side but are opposite each other are **vertical angles.** Vertical angles are equal. $\angle c$ and $\angle e$ are vertical angles.

When two lines intersect in this way, you need the measure of only one of the angles to find the measure of the others.

Example: If $\angle c = 50°$, what is the measure of each of the other angles?

Since $\angle c$ and $\angle d$ are supplementary, $\angle d$ is $180° - 50° = \mathbf{130°}$.
Since $\angle c$ and $\angle f$ are also supplementary, $\angle f$ is $180° - 50° = \mathbf{130°}$.
Since $\angle e$ is supplementary to both $\angle d$ and $\angle f$, $\angle e$ is $180° - 130° = \mathbf{50°}$.

A. Find the complement to each of the following angles.

1. 34°   85°   62°   9°   45°

B. Find the supplement to each of the following angles.

2. 45°   90°   115°   165°   122°

C. Use the figure at the right to answer the following.

3. Name two pairs of vertical angles.

4. Name two angles that are supplementary to ∠y.

5. If ∠z measures 110°, what does ∠w measure?

6. ∠z = 110°. What does ∠x measure? How do you know?

D. Use the figure at the right to answer the following.

7. Name a supplementary angle to ∠MON.

8. What is the measure of ∠POQ?

9. How many angles can you find in the figure? Name them.

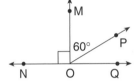

Making Connections: City Planning

Traffic management is an important part of city planning. In Midtown, Main Street and State Street are parallel for a two-mile stretch. The city planning office would like to build a road connecting the two streets within the area shown on the diagram to the right.

Your job is to propose the angle of the connecting road. Keep in mind that on a busy street, a 90° turn often slows traffic and requires special traffic signals.

1. Draw the roadway that you believe would be the most practical. Think about the flow of traffic. Explain the advantages of the road you are proposing. Estimate the measure of the angles you have created.

2. Are the angles you have created complementary, supplementary, or equal?

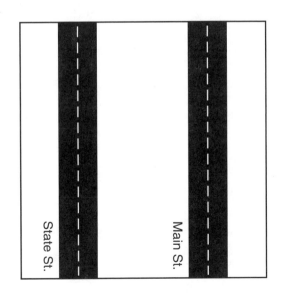

Circles

A **circle** is a closed set of points in a plane. In this case, *closed* means the ends of the circle meet. *Plane* means flat. Every point on a circle is the same distance from the center.

The distance around a circle is called the **circumference (C).** A line segment reaching from side to side of a circle and passing through the center is the **diameter (d).** Half of the diameter is the **radius (r).** The radius is the distance from the center to any point on the circle.

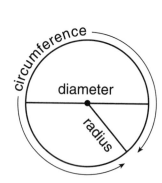

Greek mathematicians discovered a relationship between the circumference of a circle and its diameter. Whatever the size of the circle, the circumference is always a little more than three times the diameter. They called this number π (the Greek letter **pi**). They found the value of π is approximately $\frac{22}{7}$, which is close to 3.14.

It is much easier to measure diameter than circumference. Once you know the diameter of a circle, you can use a formula to find the circumference.

▶ Use this formula to find the circumference (C) of a circle:
$C = \pi d$, where $\pi \approx 3.14$ and d = diameter.

A. Use the circle to the right to solve the problems below. The letter *O* marks the center point of the circle.

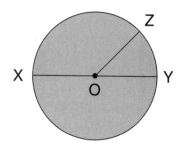

1. Line segment XY is the _____ of the circle.

2. Line segment OZ is the _____ of the circle.

3. What can you tell about the relationship between OX and OZ?

4. If OZ is 7 inches, what is the measure of XY?

5. If XY is 30 centimeters, what is the measure of OZ?

 B. Use the circumference formula ($C = \pi d$) to solve problems 6–9. You may use a calculator.

6. A metal strip is clamped around the lens of a magnifying glass. The metal strip is 1 inch longer than the circumference of the lens. If the diameter of the lens is 3 inches, what is the length of the metal strip to the nearest tenth of an inch?

190

7. Cans of frozen juice concentrate often have plastic strips that you pull to remove the lid from the can. The length of the strip is the circumference of the lid plus $\frac{2}{8}$ inch (0.25 inch). If the radius of a lid is 1.25 inches, what is the length of the plastic strip to the nearest tenth of an inch?

8. Max is planting a circular flower garden in the center of a lawn. He has 30 feet of fencing. To the nearest tenth of a foot, what is the greatest diameter the garden can have without Max buying more fencing?

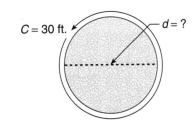

9. To find the perimeter of the figure at the right, you need to find the circumference of the circular portion of the figure. If the circular part is $\frac{3}{4}$ of a circle, what is the perimeter of the figure to the nearest tenth of a centimeter?

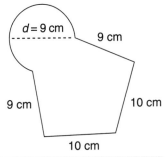

Making Connections: Partitioning

You are **partitioning** whenever you cut up something using line segments. Cutting a birthday cake is a good example of partitioning.

This birthday cake has been cut into six equal portions using three straight lines. The lines all pass through the center of the circle.

In the problems below, the lines do not have to pass through the center and the pieces do not have to be the same size. But the lines do have to be straight.

Discuss How would you solve the following problems? What answers did you get?

1. Cut the small circle at the right into as many pieces as possible using 3 straight lines. How many pieces did you get?

2. What is the minimum number of lines you will need to cut the large circle into 11 pieces?

Quadrilaterals

A **polygon** is a closed, plane (flat) figure made up of line segments. In this case, *closed* means that the sides meet.

closed   not closed

A **quadrilateral** is a polygon with four sides. Squares and rectangles are the two most common quadrilaterals.

Look at the quadrilateral EFGH shown here. Each time two sides meet, an angle is formed. The point where the sides meet is called a **vertex.** The four labeled points are the **vertices** of this figure.

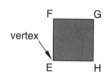

Sides EF and GH are **opposite** (across from) each other. Sides FG and EH are also opposites. Sides EF and EH are **adjacent.** This means they share an endpoint.

A straight line connecting opposite vertices (corners) is called a **diagonal.** Line segment EG is a diagonal; so is FH.

Below are examples and properties of five common quadrilaterals.

Name	Properties	Examples
square	Opposite sides are parallel. All sides are equal. All angles are right angles (90°). Diagonals are equal and perpendicular.	
rectangle	Opposite sides are equal and parallel. All angles are right angles (90°). Diagonals are equal but not necessarily perpendicular.	
rhombus	Opposite sides are equal and parallel. All sides are equal. Diagonals are perpendicular but not necessarily equal.	
parallelogram	Opposite sides are equal and parallel.	
trapezoid	Only *one* pair of sides is parallel.	

A. Use figure ABDC to answer the questions below.

1. Which side is opposite AC?

2. Name two sides that are adjacent to BD.

3. Name the two diagonals.

4. Which angle is opposite ∠CAB?

B. Use the properties of quadrilaterals to answer each question.

5. Name three properties that rectangles and squares have in common.

6. **True or False** A square, a rectangle, and a rhombus are three examples of parallelograms. Explain your answer.

7. A four-sided figure has equal and parallel opposite sides. None of the four angles are right angles, but all four sides are equal in length. What is the name of the figure?

8. **True or False** A trapezoid can *never* have a right angle. Explain your answer.

Trapezoid?

C. Imagine you are viewing the quadrilateral below from an angle that distorts the figure. You know that WX and YZ are equal, XY and WZ are equal, XW and WZ are equal, and ∠XWZ is a right angle. Use the information to answer the questions below.

9. Is it possible for this figure to have only one right angle? Why or why not?

10. What is the name for this figure?

11. **Explore** Tiles are often made in the shape of squares because it is easy to fit squares together to cover a surface. Would it be possible to use rhombus-shaped tiles to cover a surface? How about trapezoid tiles?

Triangles

A **triangle** is a polygon with three sides.

▶ The sum of the angles in a triangle is 180°.

To see this, draw a diagonal through a square. The square is divided into two equal triangles. Notice that the diagonal divides two of the right angles in half. The sum of the three angles in either of the triangles is 45° + 45° + 90° = 180°.

▶ Equal sides are opposite equal angles.

In the triangle ABC, ∠A and ∠C are equal. Both measure 45°. The sides across from them, AB and BC, are also equal.

▶ The longest side is opposite the largest angle.

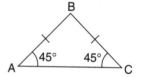

In triangle ABC, ∠B is the largest angle. The side across from it, side AC, is the longest side.

The names of the triangles come from the relationships among the angles and the sides.

Name	Properties	Examples
equilateral	three equal sides three equal angles	2 cm / 60° \ 2 cm 60° 60° 2 cm
isosceles	two equal sides two equal angles	2 cm 2 cm 70°→ ←70°
right	one right angle	
scalene	no equal sides no equal angles	2.1 cm 55° 25° 100° 1 cm 2 cm

A. Identify each triangle.

1.

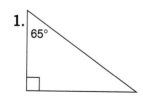

2.

3 cm / 3 cm
3 cm

3. 78°→ ←78°

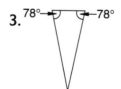

4.

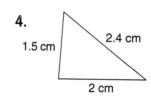

B. Use your knowledge of triangles to solve problems 5–7.

5. Determine which groups of three angles could form triangles.

    (1) 30°, 30°, 90°

    (2) 50°, 80°, 50°

    (3) 45°, 60°, 75°

    (4) 120°, 30°, 45°

    (5) 50°, 60°, 70°

6. What is the longest side in triangle CDE? Why?

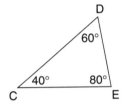

7. What is the largest angle in triangle MNO? How do you know?

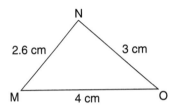

C. Use algebra to solve for the unknown angle(s).

8. What is the measure of ∠U?

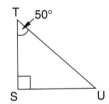

10. In the isosceles triangle VWX, ∠W measures 50°. What are the measures of ∠V and ∠X?

9. ∠C is twice the measure of ∠A. If ∠B measures 120°, what are the measures of ∠A and ∠C?

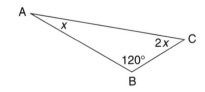

11. **Investigate** The rectangle below is divided into two triangles. Based on your knowledge of triangles, what is the sum of the angles in a rectangle?

D. Answer these questions about isosceles triangles.

12. Triangle BCD is an isosceles triangle. If sides BC and CD are equal in length, then ∠B must be equal to which other angle?

13. **Draw** Can there be a right angle in an isosceles triangle? Create a drawing that proves your answer.

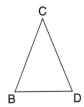

Similar Geometric Figures

Similar figures have the same shape but are different sizes. Imagine using a copy machine to reduce a figure you have drawn. The figure on the copy would be in a smaller scale than the original, but the shape would be the same.

The triangles to the right are similar because the ratios of the lengths of corresponding sides are equal. Each side in the second triangle is half as long as the first.

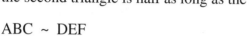

ABC ~ DEF
The symbol ~ means "is similar to."

The rectangles shown are not similar.

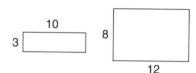

You can use proportion to solve problems about similar figures.

Finding a Missing Side

Example: The plots of land shown here are similar. Find the width of the smaller plot.

Step 1. Set up a proportion with corresponding sides.

$$\frac{\text{length}}{\text{width}} \frac{50}{40} = \frac{30}{x}$$

Step 2. Cross multiply and solve for x.
The width of the smaller plot is **24**.

$$50x = 1{,}200$$
$$x = 24$$

When the sun casts the shadow of an object perpendicular to the ground, a right triangle is formed. The height of the object is one leg of the triangle. The length of the shadow is the other leg. The longest side (called the **hypotenuse**) is the imaginary line from the top of the object to the end of the shadow.

Since the angle of the sun will be the same for two objects, you can find the height of a taller object by using a smaller object and comparing similar triangles.

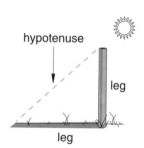

Finding the Height of a Tall Object

Example: A street sign 6 feet high casts a shadow of 4 feet. At the same time, a nearby tree casts a 10-foot shadow. How tall is the tree?

Step 1. Set up a proportion with corresponding sides.

$$\frac{\text{height}}{\text{shadow}} \frac{6}{4} = \frac{x}{10}$$

Step 2. Cross multiply and solve for x.
The height of the tree is **15 feet**.

$$4x = 60$$
$$x = 15$$

A. Decide whether each pair of figures is similar. Do not rely on the appearance of the drawings.

1.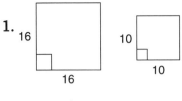

2.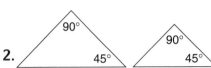

3.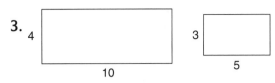

4.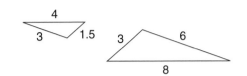

 B. Solve the following. You may use a calculator.

5. Triangles ABC and DEF are similar. Find the missing lengths.

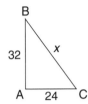

6. Karlene is using a photocopier to enlarge a graph that is 3 inches wide and 5 inches long. If her boss wants the length of the enlargement to be $7\frac{1}{2}$ inches, how wide will the graph be?

7. A billboard casts a shadow of 39 feet when a nearby 4-foot post casts a shadow of 6 feet. Find the height of the billboard.

8. On a map, a rectangular park is 1.75 centimeters wide by 2.5 centimeters long. If the park is actually 200 yards long, what is its width in yards?

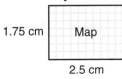 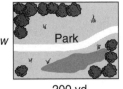

9. Triangles XYZ and XVW are similar. What is the length of VW?

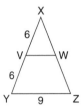

10. A building casts a shadow of 65 feet while a nearby street sign casts a shadow of 5 feet. If the street sign is 6 feet in height, how high is the building?

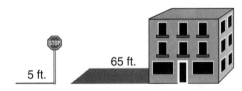

11. **Analyze** Write *true* or *false* after the following statements. Write one sentence explaining your reasoning for each one.
    a. All equilateral triangles are similar.
    b. All squares are similar.
    c. All rectangles have equal angles; therefore, all rectangles are similar.
    d. All right triangles are similar.

Finding Patterns in Algebra and Geometry

Finding a pattern is a useful problem-solving strategy. It can save time you might otherwise spend in lengthy calculations. In this lesson, you will see how patterns can be helpful in solving problems.

One example of patterns in mathematics is the number series. A **number series** is a set of numbers that continues according to a rule.

The counting numbers (1, 2, 3, . . .) are a number series in which each number is 1 larger than the number before it. In the series 1, 4, 7, 10, . . . , each number is 3 more than the preceding number.

A. **Find the next number in each of the following series. Write the rule that determines which number comes next.**

1. 3, 6, 9, 12, 15, _____

2. 100, 91, 82, 73, 64, _____

3. 1, 10, 100, 1,000, _____

4. 2, 4, 8, 16, _____

5. 2, 5, 11, 23, 47, _____

6. $1, \frac{1}{2}, \frac{1}{4}, \frac{1}{8},$ _____

B. **Complete each series.**

7.

2:00 AM PM   3:15 AM PM   4:30 AM PM   5:45 AM PM   AM PM

8. A, E, I, M, Q, _____
   (*Hint:* Think about the position of letter in the alphabet.)

You have seen how you can use proportion to solve problems that involve similar figures. Finding a pattern can often get the job done just as well. In the next example, the information is organized in a table. Tables and charts can make patterns easier to recognize.

Finding a Pattern with Similar Figures

Example: The triangles are similar. Find the measure of the missing sides.

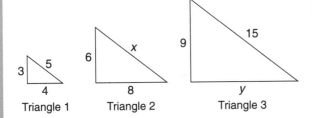

	Vertical Leg	Horizontal Leg	Hypotenuse
Triangle 1	3	4	5
Triangle 2	6	8	*x*
Triangle 3	9	*y*	15

Read down each column to find the pattern. You can quickly see that the sides of Triangles 2 and 3 are multiples of Triangle 1. The missing side in Triangle 2 is **10.** The missing side in Triangle 3 is **12.**

C. This table shows the relationship between the length and width of several rectangles and their perimeters. Find the pattern and complete the table. Then solve problems 9–10.

Rectangle	Length	Width	Perimeter
ABCD	2	1	6
EFGH	3	2	10
IJKL	4	3	14
MNOP	5	4	
QRST			
UVWX			

9. Find the perimeter of a rectangle with a length of 6 and a width of 5.

10. Find the length of a rectangle with a width of 7 and a perimeter of 30.

D. The stair steps below are built with cubes. Count the number of cubes in each set to complete the chart. Then solve the problems.

Number of steps	2	3	4	5	6
Number of cubes	3	6			

11. How many cubes are needed to make a stair step with 8 steps?

12. **Explain** How many blocks would you expect to find on the bottom row of a stair step with 20 steps? How do you know your answer is correct?

The Pythagorean Theorem

The two sides of a right triangle that form the right angle are called **legs.** The side opposite the right angle is called the **hypotenuse.** Since the right angle is the largest angle, the hypotenuse is the longest side.

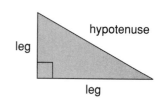

A philosopher named Pythagoras lived in Greece about 2,500 years ago. He is credited with discovering many properties about numbers, including the relationship among the sides of a right triangle.

The **Pythagorean theorem** states that the sum of the squares of the legs of a right triangle equals the square of the hypotenuse.

▶ The formula for the Pythagorean theorem is $a^2 + b^2 = c^2$ where a and b are the legs of a right triangle and c is the hypotenuse.

The illustration shows a right triangle with legs of 3 and 4 and a hypotenuse of 5.

The sum of the squares of the legs is $3^2 + 4^2 = 9 + 16 = 25$.

The square of the hypotenuse is $5^2 = 25$. $\qquad 3^2 + 4^2 = 5^2$

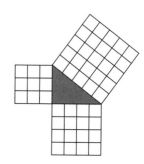

Finding the Hypotenuse

Example: The lengths of the legs of a right triangle are 9 and 12. Find the length of the hypotenuse.

Step 1. Substitute 9 for a and 12 for b in the Pythagorean theorem formula.
Step 2. Evaluate the left side of the equation.

$$a^2 + b^2 = c^2$$
$$9^2 + 12^2 = c^2$$
$$\sqrt{81 + 144} = c$$
$$\sqrt{225} = c$$

Step 3. Find c.

$$15 = c$$

The hypotenuse is **15 units** long.

Finding a Leg of a Right Triangle

Example: The hypotenuse of a right triangle has a length of 10 units, and one of the legs is 6 units long. Find the measure of the other leg.

Step 1. Substitute 10 for c and 6 for a in the Pythagorean theorem formula.
Step 2. Simplify the equation.

$$a^2 + b^2 = c^2$$
$$6^2 + b^2 = 10^2$$
$$36 + b^2 = 100$$
$$b = \sqrt{64}$$

Step 3. Find b.

$$b = 8$$

The other leg is **8 units** long.

A. Solve each problem. You may use a calculator.

1. Find the measurement of side AB.

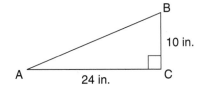

2. What is the measurement of side XZ?

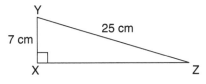

3. In the rectangle shown here, side EF is 9 centimeters and side FH is 40 centimeters. Find the measure of the diagonal FG.

4. For a circus act, an 18-foot pole is braced with wires that extend from the top of the pole to a stake in the ground 80 feet from the base of the pole. How long is each bracing wire?

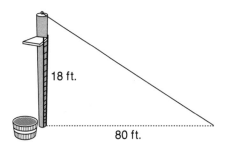

> **For another look**
> at perfect squares, turn to page 270.

B. Use the diagram to answer the questions.

5. The plans for a tiled pool call for a diamond shape in the center of the pool floor. A contractor divided the floor area into 4 rectangles with equal dimensions and then formed the diagonals as shown in the diagram. The shaded portion of the diagram will be the blue tiles, and the remaining part will be yellow. How could you find the length of the diagonals, which will be covered by border tile separating the blue and yellow tiles? What measurements do you need? Explain how you would find the answer.

6. **Analyze** How do you think the surface area of the blue-tiled portion will compare to the surface area of the yellow portion? How do you know your answer is correct?

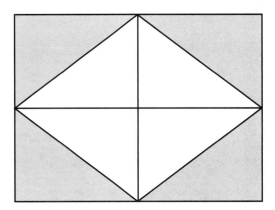

Reading Maps

Map reading involves solving proportions and occasionally applying the Pythagorean theorem. When reading any map, look for the **key** or **legend.** This usually appears in one corner, and it tells you the scale at which the map is drawn.

A. In the map below, each inch represents 8 miles. Use the map and a ruler to solve problems 1–3.

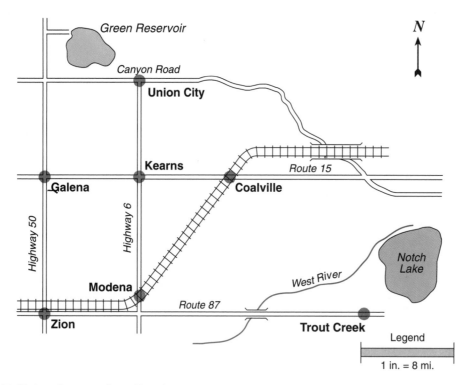

1. Tell the distance in miles from:

   a. Galena to Coalville by Route 15
   b. Kearns to Union City by Highway 6
   c. Trout Creek to Zion by Route 87
   d. Kearns through Galena to the turnoff to Green Reservoir
   e. Trout Creek to Union City by Route 87 and Highway 6

2. The distance on the map from Modena to Kearns is $1\frac{1}{4}$ inches. The distance from Kearns to Coalville is 1 inch. Find the actual distance from Modena to Coalville along the railroad tracks.

3. If a direct route were built from Trout Creek to Union City, how many miles long would the road be? Would the road pass directly through Coalville?

This map shows a section of a large apartment complex in Los Angeles. The map shows far more detail than a map that shows distances of many miles.

B. In the map below, every inch represents 300 feet. Use the map and a ruler to solve problems 4–7.

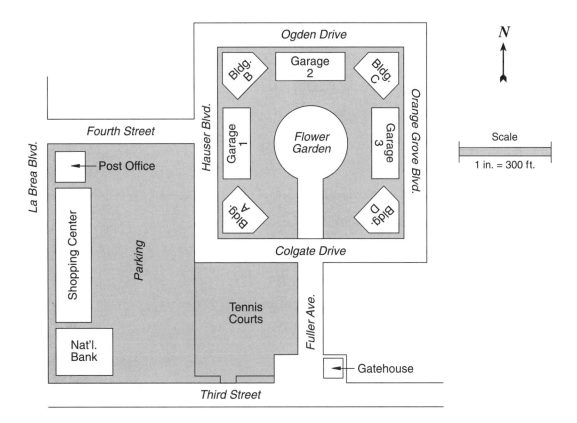

4. What is the approximate distance from Building B to the Gatehouse taking Hauser, Colgate, and Fuller?

   (1) 900 feet

   (2) 1,100 feet

   (3) 1,300 feet

   (4) 1,500 feet

5. To the nearest 10 feet, calculate the diagonal distance from the northeast corner of Hauser and Colgate to the southwest corner of Ogden and Orange Grove.

6. Fran stops at the bank and then drives to Garage 3. If she took Third Street to Fuller to Colgate to Orange Grove, approximately how far did she travel?

   (1) 1,500 feet

   (2) 1,800 feet

   (3) 2,100 feet

   (4) 2,400 feet

7. **Write** Imagine you are stopped by a driver near the corner of Ogden and Orange Grove. The car is facing north on Orange Grove. The driver asks you for directions to the tennis courts. Write the directions you would give. Be as specific as possible.

Perimeter and Circumference

You know that perimeter and circumference both show the distance around a figure.

The perimeter of the six-sided figure below is $2.5 + 2.5 + 3 + 4 + 5.5 + 6.5 = 24$.

Typical units of measure for perimeter and circumference are inches, feet, yards, miles, centimeters, meters, and kilometers.

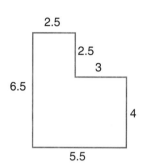

Formulas for Perimeter (P) and Circumference (C)

   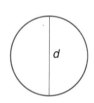 

Rectangle
$P = 2l + 2w$
where
l = length
w = width

Square
$P = 4s$
where s = side

Polygon with n sides
$P = s_1 + s_2 + \ldots + s_n$
where s_1, etc. = each side

Circle
$C = \pi d$ or $C = 2\pi r$
where
$\pi \approx 3.14$
d = diameter
r = radius

 A. Find the perimeter or circumference of each figure. You may use a calculator.

1.

$s = 15$ m

3.
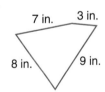
7 in.   3 in.
8 in.   9 in.

5.

3 in.   3 in.
5 in.

2.

$d = 80$ ft.

4.
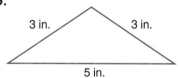
$r = 4$ cm

6.
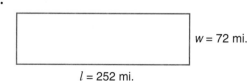
$w = 72$ mi.
$l = 252$ mi.

Sometimes you need to calculate the measure of one or more sides before you can find the perimeter.

Finding Unknown Sides

Example: What is the perimeter of the dollhouse room shown here?

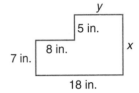

Step 1. The measure of x is the sum of the two vertical sides. Add 5 and 7 to find x.

$5 + 7 = x$
$12 = x$

Step 2. The sum of y and the 8-inch horizontal side is 18. To find y, subtract 8 from 18.

$18 - 8 = y$
$10 = y$

Step 3. Add all the sides to find the perimeter.

$10 + 12 + 18 + 7 + 8 + 5 = 60$

The perimeter of the room is **60 inches.**

 B. Find the perimeter of the following figures. You may use a calculator.

7.

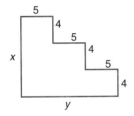

8.

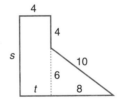

9.

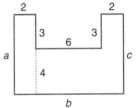

 C. Solve each problem. You may use a calculator.

10. The fountain in the diagram is enclosed by a low stone wall. What is the length of the wall?

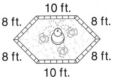

11. What is the perimeter of the swimming pool shown below? (*Hint:* Think of the pool as a rectangle with a half circle at each end.)

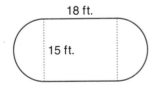

12. A builder wants to put a tile border around a large circular pool. The diameter of the pool is 9 feet. To the nearest tenth of a foot, what is the circumference of the pool?

13. Explain The stage area shown below consists of two rectangular platforms and a triangular one. Suppose you need to find the perimeter of the stage. Explain how you could find the measures of sides a and b.

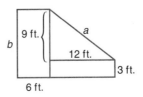

Area

To paint a wall, carpet a room, or tile a floor, you first need to measure or estimate the size of the surface you want to cover. You need to measure area. **Area** is a measure of the amount of surface.

Area is measured in **square units.** If the sides of a figure are measured in inches, its area would be measured in square inches (in.^{2} or sq. in.).

Here are several formulas for finding the areas of common figures.

Area Formulas

Rectangle	Square	Parallelogram	Triangle	Circle
$A = lw$	$A = s^2$	$A = bh$	$A = \frac{1}{2}bh$	$A = \pi r^2$
where	where s = side	where	where	where
l = length		b = base	b = base	$\pi \approx 3.14$
w = width		h = height	h = height	r = radius

Look at the examples of a parallelogram and a triangle in the table. The numbers that are multiplied together (the *base* and the *height*) must be *perpendicular* (at right angles) to each other.

 A. Find the area of each figure. Remember to write your answers in square units. You may use a calculator.

1. h = 2 in.   b = 2.75 in.

2.  7 ft.   4 ft.

3. 5 cm

4. s = 13 ft.

5. 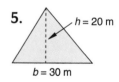 h = 20 m   b = 30 m

 B. Answer the following. You may use a calculator.

6. Martin is tiling his kitchen floor. The rectangular room measures 15 feet by 12 feet. What is the area of the room?

7. Sasha needs to buy one packet of flower seeds for every 12 square feet in her garden. If the circular garden has a diameter of 18 feet, what is the area? (*Hint:* The radius is one-half the diameter.)

In some area problems, you need to add the areas of two figures; in others, you need to find the difference between the areas of two figures.

Solving Complex Area Problems

Example 1: What is the area of the figure in this illustration?

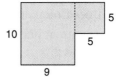

Step 1. Find the area of the rectangle.	$A = lw$	$A = 9 \cdot 10 = 90$
Step 2. Find the area of the square.	$A = s^2$	$A = 5^2 = 25$
Step 3. Add the areas.		$90 + 25 = 115$

The area of the figure is **115 square units.**

Example 2: What is the area of the shaded part of the figure?

The area is equal to the area of the rectangle minus the area of the square.

Step 1. Find the area of the rectangle.	$A = lw$	$A = 9 \cdot 10 = 90$
Step 2. Find the area of the square.	$A = s^2$	$A = 5^2 = 25$
Step 3. Subtract to find the difference.		$90 - 25 = 65$

The area of the shaded part of the figure is **65 square units.**

C. Find the area of each figure. You may use a calculator.

8.
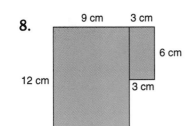

9.
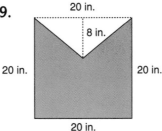

10.
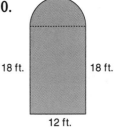

D. Solve each problem.

11. The shaded area in the figure below represents a walkway around a triangular garden. What is the area of the walkway?

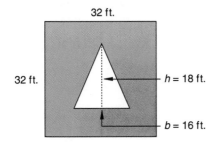

12. **Predict** Two rectangles have different dimensions but the same area. Do the rectangles have the same perimeter? Test your prediction. List the dimensions of all possible rectangles with an area of 24 square feet. (Use only whole numbers.) Now find the perimeter of each rectangle. Use this experiment to prove or disprove your prediction.

Volume

A line has only one dimension—length. The polygons and circles that you worked with in earlier lessons had two dimensions—length and width.

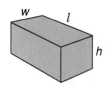

The box in the illustration is an example of a three-dimensional figure. A three-dimensional figure has length, width, and height (or depth). The shape shown here is a **rectangular solid.**

A rectangular solid has six **faces.** The faces are the rectangles or squares that make up the sides of the figure.

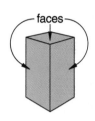

The space inside a three-dimensional solid is called **volume.** Volume is a measure of how much a solid can hold. For example, volume tells how much water a swimming pool holds or how much air is in a room.

Volume is measured in **cubic units.** Any unit of length can be cubed. Some common units of measure are cubic inches (in.^{3}), cubic feet (ft.^{3}), cubic yards (yd.^{3}), cubic centimeters (cm^3), and cubic meters (m^3).

A cubic foot, for example, is one foot long, one foot wide, and one foot high. When you measure volume in cubic feet, you are finding out how many cubic feet it takes to fill up the space.

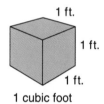

1 cubic foot

Here are several helpful formulas for finding volume.

Volume Formulas

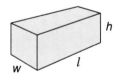

Rectangular Solid
$V = lwh$
where l = length
w = width
h = height

Cube
$V = s^3$
where s = side

Cylinder
$V = \pi r^2 h$
where $\pi \approx 3.14$
r = radius
h = height

A. Find the volume of each figure. Use 3.14 for π and round your answer to the nearest tenth if necessary. You may use a calculator.

1.

2 in. 8 in.
6 in.

3.

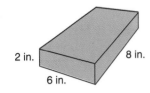

r = 3 cm
7 cm

5.

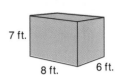

7 ft.
8 ft. 6 ft.

2.

s = 3 cm

4.

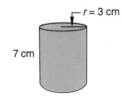

r = 14 in.
40 in.

6.

10 ft.
20 ft. 1½ ft.

B. Answer the following. Make sure you write your answer in cubic units. You may use a calculator.

7. A shipping container is 6 feet by 4 feet by 3 feet. What is its volume?

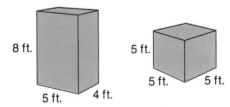

6 ft.
3 ft.
4 ft. THIS END UP

8. How much greater is the volume of the first box than the volume of the second?

8 ft. 5 ft.
5 ft. 5 ft.
5 ft. 4 ft.

9. A cylindrical fuel tank has a height of 20 feet and a diameter of 16 feet. What is its volume in cubic feet?

10. A plan for a concrete wall calls for it to be ½ foot thick, 6 feet high, and 48 feet long.

   a. What is the volume of the wall in cubic feet?

   b. Concrete is sold in cubic yards. There are 27 cubic feet in a cubic yard. How many cubic yards of concrete are needed to build the wall?

11. **Explore** Suppose you have two rectangular pieces of paper, each 6 inches by 8 inches. The pieces are rolled to make different cylinders. The diagram below shows the height and radius for each cylinder. You know the pieces of paper had equal surface areas. Do the cylinders formed from the pieces have equal volume?

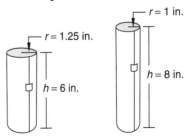

r = 1 in.
r = 1.25 in.
h = 8 in.
h = 6 in.

Choosing Area, Perimeter, or Volume

You know how to find perimeter, area, and volume of various figures. But knowing which to solve for takes practice.

Remember that **perimeter** measures the distance around a figure, **area** measures the surface of a two-dimensional figure, and **volume** measures the space inside a three-dimensional figure.

A. For each problem, decide whether to solve for perimeter, area, or volume. Then solve the problem. You may use a calculator.

1. Marisol wants to put a satin border on a quilt she has made. The illustration shows the dimensions of the quilt. What will be the length of the border?

   a. Perimeter, area, or volume?
   b. Answer:

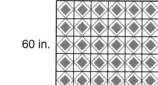

2. Max needs to know how much the moving truck shown in the ad will hold. How much room is there inside it?

   a. Perimeter, area, or volume?
   b. Answer:

3. A revolving sprinkler sprays a lawn for a distance of 18 feet. How much ground is covered by the sprinkler as it makes one complete revolution?

   a. Perimeter, area, or volume?
   b. Answer:

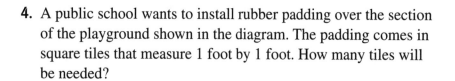

4. A public school wants to install rubber padding over the section of the playground shown in the diagram. The padding comes in square tiles that measure 1 foot by 1 foot. How many tiles will be needed?

   a. Perimeter, area, or volume?
   b. Answer:

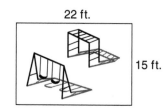

B. Write the correct unit of measure for each situation. Then use the information from the illustration to solve the problem. You may use a calculator.

5. Find how much chrome edging is needed to finish a rectangular kitchen table.

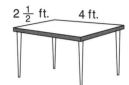

    a. Unit of measure:

    b. Solution:

6. How much water will a circular wading pool hold?

    a. Unit of measure:

    b. Solution:

7. Find the surface area of the face of the triangular wood brace shown here.

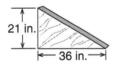

    a. Unit of measure:

    b. Solution:

> **For another look**
> at formulas, turn to
> page 271.

Making Connections: Estimating with Pi (π)

You may be surprised to learn that when you perform a calculation with π, you are actually estimating. In the exact value of π, the decimal continues endlessly. We round the value to 3.14. But you can simplify calculations even more by rounding π to 3.

Choose the correct answer for each problem. Make your work easier by using 3 for π.

1. Find the circumference of a circle when its diameter is 8 feet. (*Hint: C = πd*)

    (1) 12.56 feet

    (2) 25.12 feet

    (3) 33.04 feet

2. Find the area of a circle with a radius of 14 inches. (*Hint: A = πr²*)

    (1) 153.86 square inches

    (2) 419.44 square inches

    (3) 615.44 square inches

3. Find the volume of a storage tank with a height of 25 meters. The radius of the circular base is 10 meters. (*Hint: V = πr²h*)

    (1) 1,962.5 cubic meters

    (2) 7,850 cubic meters

    (3) 19,625 cubic meters

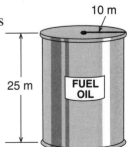

The Coordinate System

How do you give directions? One way is to describe the path the person should take. "Drive three blocks north and six blocks west." Another way is to name major cross streets. "City Hall is near the corner of Fifth Street and Lincoln Boulevard." Both of these methods work because many cities are organized in a grid of intersecting streets.

A **rectangular coordinate system** functions in the same way. A coordinate system is made up of two perpendicular lines that intersect at zero. The horizontal line is the **x-axis.** The vertical line is the **y-axis.** The point where the lines intersect is the **origin.**

Every point on the system represents an **ordered pair** of numbers (x,y). The x-coordinate, which always comes first, tells how far a point lies to the left or right of the origin. The y-coordinate tells how far a point lies above or below the origin. The x-axis and the y-axis divide the system into four sections called **quadrants.**

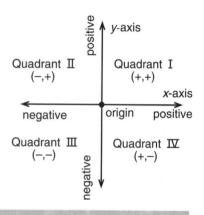

Identifying Points on the Coordinate System

Example 1: What are the coordinates of Point B? B is 4 units right of the origin and 1 unit below it. The coordinates are **(4,–1).**

Example 2: What are the coordinates of Point C? C is 2 units left of the origin and 3 units above it. The coordinates are **(–2,3).**

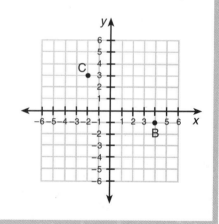

A. Write the coordinates for the points on this rectangular coordinate system.

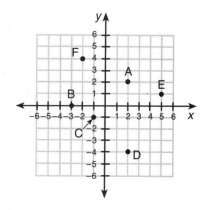

1. A =   3. C =   5. E =

2. B =   4. D =   6. F =

212

B. Plot the points described below on this rectangular coordinate system. Be sure to write the letter label by each point.

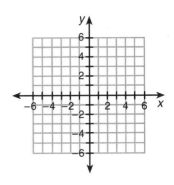

7. G = (4,–1) **9.** I = (0,3) **11.** K = (2,2)

8. H = (–1,5) **10.** J = (–4,0) **12.** L = (–2,–5)

Finding the Distance between Two Points

Example 1: What is the distance from M to N?
Find the coordinates for both points: M = (–1,–1) and N = (3,–1).
The *y*-value is the same for both. Subtract the *x*-values:
–1 – 3 = –4 or 3 – (–1) = 4. Ignore the sign of the difference.
The points are **4 units** apart.

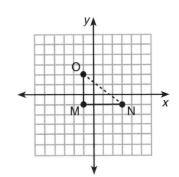

Example 2: What is the distance from M to O?
Find the coordinates: M = (–1,–1) and O = (–1,2). Subtract
the *y*-values: –1 – 2 = –3 or 2 – (–1) = 3. Ignore the sign.
The distance is **3 units.**

Example 3: Find the distance between O and N.

Segment MO and segment MN form a right angle. Segment
ON forms the hypotenuse of a right triangle. You can use the
Pythagorean theorem to find the distance between the points.*
The distance between O and N is **5 units.**

$$a^2 + b^2 = c^2$$
$$\sqrt{a^2 + b^2} = c$$
$$\sqrt{3^2 + 4^2} = c$$
$$\sqrt{9 + 16} = c$$
$$\sqrt{25} = c$$
$$5 = c$$

---

*To review the Pythagorean theorem, turn to page 200.

C. Use figure PQSR to answer these questions.

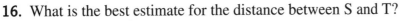

13. What is the distance from R to S?

14. What is the distance between S and Q?

15. Find the distance between Q and R.

16. What is the best estimate for the distance between S and T?

    (1) between 5 and 6 units   **(2)** between 6 and 7 units   **(3)** between 7 and 8 units

17. **Explain** How could you find the area of the figure formed by points P, Q, S, and R?
    Assume each square on the grid represents 1 square inch. Find the area of PQSR.

Slope and Intercept

Slope refers to the amount of incline, slant, or steepness. Ski runs, most roofs, and the graphs of linear equations have slope.

Slope is the ratio of the **rise** to the **run** of an incline. The slope of the ramp shown here is 1:3. This means that for every foot of vertical distance (rise), there are 3 feet of horizontal distance (run).

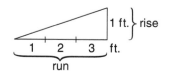

To calculate the slope of a line on the coordinate system, use this formula:

slope $(m) = \frac{y_2 - y_1}{x_2 - x_1}$ where (x_1, y_1) and (x_2, y_2) are two points on the coordinate system.

Calculating Slope

Example 1: What is the slope of the line that passes through points A and B?
Point A (1,1) is (x_1, y_1) and Point B (4,3) is (x_2, y_2).
Substitute these values into the formula and simplify.
$m = \frac{3 - 1}{4 - 1} = \frac{2}{3}$
The slope is $\frac{2}{3}$.

Example 2: What is the slope of the line that passes through points C and D?
Point C (–1,4) is (x_1, y_1) and Point D (2,–2) is (x_2, y_2).
Substitute and solve. Simplify.
$m = \frac{-2 - 4}{2 - (-1)} = \frac{-6}{3} = -2$
The slope is **–2.**

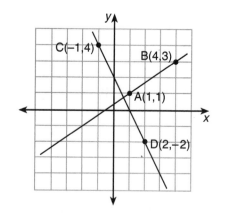

A line that *rises* from left to right has **positive slope.**
A line that *falls* from left to right has **negative slope.**

A. Calculate the slope for each line.

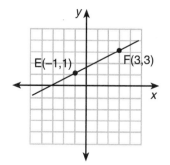

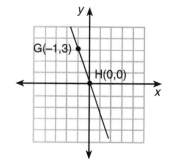

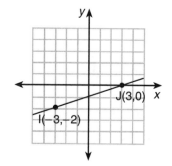

1. Slope for line EF _____

2. Slope for line GH _____

3. Slope for line IJ _____

The illustration shows the graph of the equation $y = x + 3$. The line crosses the y-axis at (0,3). This point is called the **y-intercept**. The line crosses the x-axis at (–3,0). This point is called the **x-intercept.**

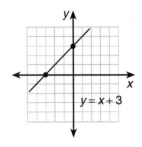

B. Use the graph at the right to answer questions 4 and 5.

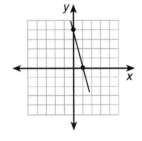

4. What are the coordinates of the y-intercept? _____

5. What are the coordinates of the x-intercept? _____

To find the y-intercept of an equation, substitute 0 for x, and solve the equation for y. To find the x-intercept of an equation, substitute 0 for y, and solve the equation for x.

Finding the x- and y-Intercepts

Example: Find the x- and y-intercepts of this equation:
$3x - 5y = 15$

Step 1. Substitute 0 for x, and solve for y.

$3(0) - 5y = 15$
$-5y = 15$
$y = \frac{15}{-5} = -3$

The y-intercept is **(0,–3)**.

Step 2. Substitute 0 for y, and solve for x.

$3x - 5(0) = 15$
$3x = 15$
$x = \frac{15}{3} = 5$

The x-intercept is **(5,0)**.

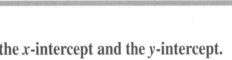

C. For each problem, calculate the coordinates of the x-intercept and the y-intercept. Plot the points on the coordinate system and draw a line through the points.

6. $2x + 3y = 6$

   y-intercept =

   x-intercept =

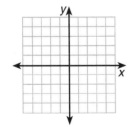

7. $x - y = 4$

   y-intercept =

   x-intercept =

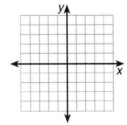

8. $3x + y = 3$

   y-intercept =

   x-intercept =

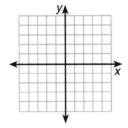

9. **Explain** Is it possible for the x- and y-intercepts to have the same coordinates? Explain why you believe your answer is correct.

Unit 5 Review

A. Evaluate the following problems.

1. $8(-4)(2) =$

2. $-15 + (-6) + 3 =$

3. $\frac{-100}{-10} =$

4. $\sqrt{169}$

5. $(10 - 6)^2 + 14 =$

6. $3(19 - 4) =$

B. Choose an equation for each situation described below.

7. Fifteen less than a number is 85.
   (1) $85 - n = 15$
   (2) $15 + 85 = n$
   (3) $n - 15 = 85$

8. The quotient of 80 divided by a number increased by 15 is 35.
   (1) $\frac{80}{n} + 15 = 35$
   (2) $n + \frac{15}{80} = 35$
   (3) $\frac{80}{15} + n = 35$

9. A rectangle has a width of x. The width times the length of 12 is 121 square units.
   (1) $2(12) + 2x = 121$
   (2) $12x = 121$
   (3) $12 + x = 121$

10. Three times a number increased by 18 is equal to 9 times that number.
   (1) $3x + 9x = 18$
   (2) $9x - 3x = 18$
   (3) $3x + 18 = 9x$

C. Solve the following equations and inequalities.

11. $8y - 7 = 7y + 3$

12. $2a + 3 > 8a - 2$

13. $10x - 5 = 8x - 25$

14. $5b - 2 \geq 3b$

15. $6(z - 5) = 3z - 26$

16. $9(5n - 3) = 7(4n + 1)$

D. Use substitution to solve the following.

17. Find the value of a if $2a^2 - b = 15$ and $b = 3$.

18. Solve for x if $5x - 4y = 31$ and $2x + y = 2$.

19. Find the value of c and d if $c - d = 4$ and $c = 2d - 2$.

20. Solve for s and t if $5s - 3t = 14$ and $t = 5s + 2$.

E. Solve the following problems. You may use a calculator.

21. What length of metal pipe do you need to make a basketball hoop with a diameter of 18 inches?

18 in.

22. Rectangle ABCD has a length of 24 centimeters and a width of 12 centimeters.

    a. ∠BAC has a measure of 65°. What is the measure of ∠CAD?

    b. Find the perimeter of ABCD.

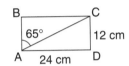

B   C
65°   12 cm
A 24 cm D

23. A tower casts a shadow 42 feet long. At the same time a 6-foot sign casts a shadow of 7 feet. How high is the tower?

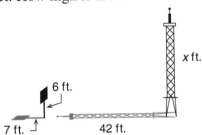

x ft.

6 ft.

7 ft.   42 ft.

24. How many cubic meters of space are in a storage area that measures 6 meters by 6 meters by 6 meters?

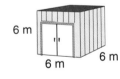

6 m
6 m   6 m

25. Triangle EFG has two equal sides and two equal angles.

    a. What kind of triangle is EFG?

    b. What is the measure of ∠F?

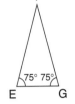

F

75° 75°
E   G

26. Find the area of the following figure. All angles are right angles.

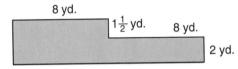

8 yd.
$1\frac{1}{2}$ yd. 8 yd.
2 yd.

27. How high up on a wall does a 50-foot ladder reach if the foot of the ladder is 14 feet from the wall?

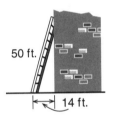

50 ft.

14 ft.

28. Find the distance between points A and B on the rectangular coordinate system.

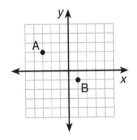

y

A
x
B

Working Together

Work in pairs or small groups. Each group should have a map. Using the legend, write an expression for calculating any distance on the map in miles. Then measure the perimeter of the area covered by the map. What is this distance in miles? Now calculate the number of square miles covered by the map.

Posttest

Solve the following problems.

Use the map to solve problem 1.

1. Denise drove from Columbus to Athens, then from Athens to Cincinnati. Finally, she drove 56 miles from Cincinnati to Dayton. How many miles did she drive?

   (1) 130 miles   **(4)** 282 miles

   (2) 208 miles   **(5)** 564 miles

   (3) 226 miles

2. The hold of a container ship has a volume of 60,000 cubic feet. If a bushel is equal to 1.25 cubic feet, how many bushels can the ship hold?

   (1) 4,800 bushels

   (2) 7,500 bushels

   (3) 48,000 bushels

   (4) 75,000 bushels

   (5) Not enough information is given.

3. Choose the equation that shows "twelve more than eight times a number is fifty-two."

   (1) $12 + 8 = \frac{52}{x}$

   (2) $8x = 12(52)$

   (3) $12(8x) = 52$

   (4) $8x + 12 = 52$

   (5) $\frac{8x}{12} = 52$

4. A plumber cut a $5\frac{3}{4}$-foot piece of water pipe from a $12\frac{1}{2}$-foot length of pipe. How many feet of pipe were left?

   (1) $6\frac{1}{4}$ feet

   (2) $6\frac{3}{4}$ feet

   (3) $7\frac{1}{4}$ feet

   (4) $7\frac{1}{2}$ feet

   (5) $8\frac{1}{4}$ feet

Problems 5 and 6 refer to the following information.

Great deals on hot new software titles!

Software Title	Regular Price	Sale Price	Member's Price
Treasure Town	$47.36	$42.12	$39.56
Reading Circus	$38.93	$36.98	$26.95
Space Wars	$31.46	$29.98	$19.95

5. Ricardo decides to buy Space Wars and Reading Circus for his children. If Ricardo pays the member's price, which of the following expressions can be used to find out how much he saves over the regular price?

   (1) ($38.93 − $36.98) + ($31.46 − $29.98)

   (2) ($31.46 − $19.95) + ($38.93 − $26.95)

   (3) ($38.93 + $26.95) − ($31.46 + $19.95)

   (4) ($31.46 + $19.95) − ($38.93 + $26.95)

   (5) ($38.93 − $26.95) − ($31.46 − $19.95)

6. The store owner decides to discount the member's price for Treasure Town an additional 25%. What will the new member's price be?

   (1) $35.52   **(4)** $11.84

   (2) $31.59   **(5)** $9.89

   (3) $29.67

7. The total attendance at Dodger Stadium for a 6-game home stand was 289,512. What was the average attendance per game?

   (1) 4,825
   (2) 28,912
   (3) 48,252
   (4) 144,756
   (5) 1,737,072

8. Marco needs at least 81 feet of fencing. The warehouse store sells fencing by the yard. How many yards of fencing does Marco need?

   (1) 2.25 yards
   (2) 6.75 yards
   (3) 27 yards
   (4) 243 yards
   (5) 972 yards

9. A car can travel 270 miles on 15 gallons of gasoline. How far could the same car travel on a full tank of 22 gallons?

   (1) 292 miles
   (2) 330 miles
   (3) 396 miles
   (4) 5,940 miles
   (5) Not enough information is given.

10. A family spends $\frac{3}{8}$ of its monthly income for rent and utilities. If the family's monthly income is $2,512, how much is spent on rent and utilities per month?

    (1) $314
    (2) $628
    (3) $670
    (4) $942
    (5) $1,256

Problems 11 and 12 refer to the graph.

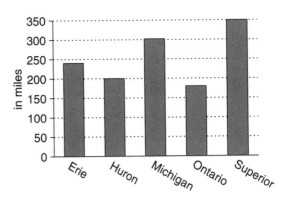

Lengths of the Great Lakes

Source: World Almanac Book of Facts, 1994

11. About how much longer is Lake Michigan than Lake Ontario?

    (1) 150 miles
    (2) 100 miles
    (3) 50 miles
    (4) 25 miles
    (5) Not enough information is given.

12. The length of which lake represents the median of the set of data?

    (1) Erie
    (2) Huron
    (3) Michigan
    (4) Ontario
    (5) Superior

Problems 13 and 14 refer to the following figure.

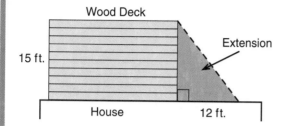

Wood Deck

15 ft.

Extension

House   12 ft.

13. Karla is adding a triangular extension to the deck shown above. Which of the following expressions can be used to find the area of the extension?

    (1) 12(15)
    (2) 2(12)(15)
    (3) $\frac{1}{2}(12 + 15)$
    (4) 2(12) + 2(15)
    (5) $\frac{1}{2}(12)(15)$

14. What is the approximate length of the longest side of the triangle?

    (1) between 17 and 18 feet
    (2) between 19 and 20 feet
    (3) between 26 and 27 feet
    (4) between 89 and 90 feet
    (5) between 143 and 144 feet

Problem 15 refers to the graph.

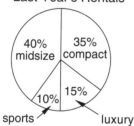

Last Year's Rentals

40% midsize   35% compact

10% sports   15% luxury

15. A car rental company needs to decide how many of each type of car to have in its fleet. It bases its decision on the previous year's rentals. Out of 200 cars, how many should be compact cars?

    (1) 10
    (2) 15
    (3) 35
    (4) 40
    (5) 70

16. A section of wire 4.25 centimeters long weighs 34 grams. What will a 60-centimeter section of the same type of wire weigh?

    (1) 98.25 grams
    (2) 255 grams
    (3) 480 grams
    (4) 2,040 grams
    (5) Not enough information is given.

17. Which of the following is the decimal equivalent for $3\frac{7}{8}$?

    (1) 3.125
    (2) 3.375
    (3) 3.625
    (4) 3.875
    (5) 3.143

18. Stuart, Brad, and Robin are marathon runners. They run three times a week to keep in shape. Last week, Robin ran twice as many miles as Stuart but 14 miles less than Brad. Together, they ran a total of 79 miles. How many miles did Brad run that week?

    (1) 65 miles
    (2) 40 miles
    (3) 39 miles
    (4) 26 miles
    (5) 13 miles

Problem 19 refers to the drawing.

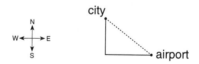

city

N
W ← → E
S

airport

19. Jamal has to drive 30 miles south and 40 miles east to get to the airport from the city. If a highway ran straight from the city to the airport, how many miles could he save on a trip to the airport?

    (1) 20 miles
    (2) 30 miles
    (3) 40 miles
    (4) 50 miles
    (5) 70 miles

Problems 20 and 21 refer to the following graph.

United States Production
Copper and Silver
1987–1992

Copper – – – –   Silver ———

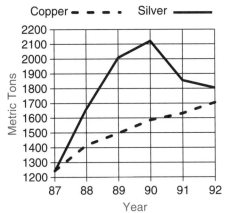

Source: World Almanac Book of Facts, 1995

20. Which of the following is a true statement about copper and silver production for the years 1987 and 1988?

   (1) Silver and copper production were roughly equal for both years.

   (2) Copper production was nearly twice that of silver production for both years.

   (3) Copper production increased significantly in 1988 while silver production remained the same.

   (4) Both silver and copper production levels rose about 1,500 metric tons in 1988.

   (5) Silver and copper production were equal in 1987, but more silver than copper was produced in 1988.

21. If the 1991–1992 trend for both silver and copper production continued, what is the most reasonable estimate of silver or copper production for 1993?

   (1) between 1,700 and 1,800 metric tons

   (2) between 1,600 and 1,700 metric tons

   (3) between 1,500 and 1,600 metric tons

   (4) between 1,400 and 1,500 metric tons

   (5) between 1,300 and 1,400 metric tons

Problem 22 refers to the drawing.

.75 ft.

6 ft.   4 ft.

22. Mona wants a cover for the sandbox to keep animals and rain out. Which expression can be used to find the amount of material needed for a cover?

   (1) $6 + 4 + .75$

   (2) $2(6 + 4)$

   (3) $6 \cdot 4$

   (4) $.75 \cdot 6$

   (5) $.75 \cdot 6 \cdot 4$

23. It took Yvonne 3 hours to travel 135 miles. Choose the expression that shows Yvonne's average speed.

   (1) $3 + 135$

   (2) $135 - 3$

   (3) $3(135)$

   (4) $\frac{3}{135}$

   (5) $\frac{135}{3}$

24. A chemist analyzes 1,200 grams of a solution and finds that 156 grams are acid. What percent of the solution is acid?

   (1) 3.6%

   (2) 7.7%

   (3) 12%

   (4) 13%

   (5) Not enough information is given.

1. **(4) 282 miles**

   Add: 74 + 152 + 56 = 282

2. **(3) 48,000 bushels**

   Divide: 60,000 ÷ 1.25 = 48,000

3. **(4)** $8x + 12 = 52$

4. **(2) $6\frac{3}{4}$ feet**

   Subtract: $12\frac{1}{2} - 5\frac{3}{4} = 12\frac{2}{4} - 5\frac{3}{4} = 11\frac{6}{4} - 5\frac{3}{4} = 6\frac{3}{4}$

5. **(2) ($31.46 – $19.95) + ($38.93 – $26.95)**

   You need to find the difference between the regular and member's prices for each software package, then add them together to find the total savings.

6. **(3) $29.67**

   25% of $39.56 is $9.89.

   Subtract the discount to find the new member's price: $39.56 – $9.89 = $29.67.

7. **(3) 48,252**

   Divide: 289,512 ÷ 6 = 48,252

8. **(3) 27 yards**

   Convert feet to yards. Divide the number of feet by three.

   81 ÷ 3 = 27

9. **(3) 396 miles**

   Set up a proportion and solve: $\frac{270 \text{ miles}}{15 \text{ gallons}} = \frac{x \text{ miles}}{22 \text{ gallons}}$

   $15x = 270(22)$

   $\frac{270\,(22)}{15} = 396$

10. **(4) $942**

    Multiply: $\$2,512 \times \frac{3}{8} = \frac{\overset{314}{\cancel{\$2,512}}}{1} \times \frac{3}{\cancel{8}} = \$942$

11. **(2) 100 miles**

    Lake Ontario is about 200 miles in length and Lake Michigan is about 300 miles. The difference is about 100 miles.

12. **(1) Erie**

    The median is the middle value in a set of numbers. In order from greatest to least, the lengths in miles would be approximately 350 (Lake Superior), 300 (Lake Michigan), 250 (Lake Erie), 200 (Lake Huron), and 180 (Lake Ontario). The middle value is 250 miles.

13. **(5) $\frac{1}{2}$(12)(15)**

    The formula for finding the area of a triangle is $A = \frac{1}{2}bh$.

14. **(2) between 19 and 20 feet**

    Since the two known sides form a right angle, you know the unknown side is the hypotenuse. Use the Pythagorean theorem: $a^2 + b^2 = c^2$

    $15^2 + 12^2 = c^2$

    $\sqrt{225 + 144} = \sqrt{369}$

    Estimate the square root of 369. You know that the square of 20 is 400. The answer must be less than 20. The square of 19 is 361. The length of the longest side must be between 19 and 20 feet.

15. **(5) 70**

    35% of 100 = 35, so

    35% of 200 = 70

    .35(200) = 70

16. **(3) 480 grams**

    Set up a proportion and solve.

    $\frac{4.25 \text{ centimeters}}{34 \text{ grams}} = \frac{60 \text{ centimeters}}{x \text{ grams}}$

    $4.25x = 34(60)$

    $4.25x = 2,040$

    $x = 480$

17. **(4) 3.875**

    Change $\frac{7}{8}$ to a decimal by dividing: 7 ÷ 8 = 0.875.

    $3\frac{7}{8} = 3.875$

18. **(2) 40 miles**

Stuart	Brad	Robin
x	$2x + 14$	$2x$

    $x + 2x + 14 + 2x = 79$

    $5x + 14 = 79$

    $5x = 65$

    $x = 13$

    Brad ran $2x + 14$ or $2(13) + 14 = 40$ miles.

19. **(1) 20 miles**

    Use the Pythagorean theorem.

    $\sqrt{30^2 + 40^2} = \sqrt{2,500}$

    $\sqrt{2,500} = 50$; total distance now: 30 + 40 = 70; distance saved: 70 – 50 = 20

20. **(5) Silver and copper production were equal in 1987, but more silver than copper was produced in 1988.**

Silver production is represented by the solid line. Both lines start at basically the same point, but the solid line crosses the 1988 scale line at a much higher point than the dotted line.

21. **(1) between 1,700 and 1,800 metric tons**

Continue the line segments at the same slope as shown between 1991 and 1992. Copper production would increase about 70 metric tons and silver production would fall about 50 metric tons. The 1993 total for both would be midway between the 1,700 and 1,800 marks on the scale.

22. **(3) $6 \cdot 4$**

Find the area. Length multiplied by width: $6 \cdot 4$.

23. **(5) $\frac{135}{3}$**

Time and distance are given. Find the rate.
$r = \frac{d}{t}$

24. **(4) 13%**

Divide: $156 \div 1,200 = 0.13 = 13\%$.

Posttest Evaluation Chart

Make note of any problems that you answered incorrectly. Review the skill area for each of those problems, using the unit number given.

Problem Number	Skill Area	Unit
1	Whole numbers	1
2, 5	Decimals and money	2
4, 10, 17	Fractions	3
6, 24	Percents	3
9, 16	Ratio and proportion	3
8, 23	Measurement	4
7, 11, 12, 15, 20, 21	Analyzing data	4
3, 18	Algebra	5
13, 14, 19, 22	Geometry	5

Answer Key

Unit 1

Addition pp. 16–17

Part A

Estimates will vary. Exact answers are given.

1. 79   78 $87 $49
2. $397 $598   759   369
3. $4,685   9,898   92 $104
4. 982   1,161 $733   1,132
5. $5,012   13,617 $6,484 $4,284

Part B

Estimates will vary.

6. **$76**

   Estimate: $50 + $30 = $80
   Exact: $48 + $28 = $76

7. **6,475 pounds**

   2,379 + 4,096 = 6,475

8. **116 home runs**

   27 + 43 + 46 = 116

9. **636 miles**

   391 + 245 = 636

10. **961 jobs**

    569 + 392 = 961

11. **$1,584**

    The two most popular booths were the Skeeball and the Dunk Tank. $649 + $935 = $1,584

12. Estimates will vary. Yes, the chairperson's prediction came true. $185, $298, and $368 round to $200, $300, and $400. $200 + $300 + $400 = $900, an amount less than the amount raised by the Dunk Tank.

Subtraction pp. 18–19

Part A

Estimates will vary. Exact answers are given.

1. 27   22 $122   242
2. $820   2,013 $1,150   9,132
3. 18   33 $784 $116
4. 112 $125 $311   939
5. $1,075   3,780 $1,590 $1,185

Part B

Estimates will vary.

6. **18 home runs**

   Estimate: 60 − 40 = 20
   Exact: 61 − 43 = 18

7. **$439**

   $824 − $385 = $439

8. **$225**

   $605 − $380 = $225

9. **923 feet deep**

   1,330 − 407 = 923

Part C

10. **637 points**

    3,041 − 2,404 = 637

11. **988 points**

    4,029 − 3,041 = 988

12. **1988, 1989, 1991, 1992**

    You may have included these ideas in your answer:

    • Jordan may have played in fewer games because of sickness or injury.
    • Defensive play may have been better in those years.
    • More shots were taken as a team instead of just Jordan carrying the team.
    • Jordan may have had less playing time.

Multiplication pp. 20–21

Part A

Estimates will vary. Exact answers are given.

1. 68   232   825   3,136   14,595

Part B

Estimates will vary. Exact answers are given.

2. 1,088 $3,168   11,043   2,625   31,122
3. $4,150   24,024   77,625   351,405   1,108,722
4. 5,625 $4,140 $6,500   251,364   338,856

Part C

Estimates will vary.

5. **836 seats**

   Estimate: 20 × 40 = 800
   Exact: 22 × 38 = 836

6. **$624**

   52 × $12 = $624

7. 240 gallons

$4 \times 12 = 48$ bottles
$48 \times 5 = 240$ gallons

8. 140 miles

$35 \times 4 = 140$

Part D

9. 1,680 customers

$420 \times 4 = 1,680$

10. 5,720 customers

$560 - 450 = 110$ more customers in 1 week
$110 \times 52 = 5,720$ customers

11. The supper shift should be cut.

The breakfast shift makes $420 \times \$3 = \$1,260$.
The lunch shift makes $560 \times \$2 = \$1,120$.
The supper shift makes $450 \times \$2 = \$900$.

Division pp. 22–23

Part A

Estimates will vary. Exact answers are given.

1. 27   198 R8   205   880
2. 2,963 R1   1,217 R1   1,516   1,994 R4
3. 379 R1   1,788 R1   515   1,157 R7
4. 1,587 R1   1,408   2,034   38,812 R4

Part B

Estimates will vary.

5. $78 per room

Estimate: $\$700 \div 10 = \$70$
Exact: $\$702 \div 9 = \$78$

6. 150 boxes

$1,200 \div 8 = 150$

7. $365

$\$1,095 \div 3 = \$365$

8. 16 temporary employees

$128 \div 8 = 16$

Making Connections: Using Compatible Numbers p. 23

You may have used different compatible numbers to make the estimate.

1. Estimate: $15,000 \div 5 = 3,000$ Exact: 2,997 R1
2. Estimate: $6,400 \div 8 = 800$ Exact: 816 R2
3. Estimate: $45,000 \div 9 = 5,000$ Exact: 5,288 R8
4. Estimate: $49,000 \div 7 = 7,000$ Exact: 6,800

Division by Two or More Digits pp. 24–25

Part A

1. 2   15   8   14   36 R36
2. 13 R24   38 R12   125   249   853 R11

Part B

3. 205   2,304   860   254 R5
4. 605 R3   8,041   80   602

Part C

Estimates will vary.

5. $408

Estimate: $\$1,600 \div 4 = \$400$
Exact: $\$1,632 \div 4 = \$408$

6. 105 centimeters

$2,100 \div 20 = 105$

7. 14 points

$252 \div 18 = 14$

8. 604 cars

$220,460 \div 365 = 604$

Part D

9. $843

$\$10,116 \div 12 = \$843$

10. $1,900

The Ruizes want to save an additional $2,000 next year. Divide $2,000 by 4 to find the amount to cut from each of the 4 categories. $2,000 \div 4 = \$500$. Subtract $500 from the amount spent on clothing: $\$2,400 - \$500 = \$1,900$.

11. Transportation

One approach is to multiply the amount for each item by 3. Then see which of these totals is nearest to $11,500, the amount spent on rent.

A faster approach is to divide $11,500 by 3. Then compare the result ($3,833) to the other amounts in the list.

You could also use estimation:
Rent = $\$11,500 \approx \$12,000$
$\$12,000 \div 3 = \$4,000$
Transportation = $\$3,888 \approx \$4,000$

Mental Math and Estimation pp. 26–27

Your methods of estimation may differ.

Part A

1. Estimate: $500 \times 20 = 10,000$
   Exact Answer: **(2)** 9,348

2. Estimate: 5,000 + 5,000 + 5,000 = 15,000
   Exact Answer: **(2)** 15,112

3. Estimate: 1,200 ÷ 60 = 20
   Exact Answer: **(3)** 22

Part B

4. **(3) between $22,000 and $24,000**
   Estimate: $2,000 × 12 = $24,000

5. **(2) $900**
   Estimate: $3,600 ÷ 4 = $900

6. **(2) 7 or 8 buses**
   Estimate: 16,000 ÷ 2,000 = 8 buses

Making Connections: When Is an Estimate Enough? p. 27

Your answers may differ. If you disagree, discuss your reasons with your teacher or a partner.

1. Exact
2. Rounded
3. Rounded
4. Exact
5. Exact
6. Rounded

The Five-Step Plan pp. 28–29

1. **a.** After spending $99 and $129, how much does Stuart have left?

   b. Needed: $310 starting budget, $99 and $129 in expenses

   c. Subtraction: $310 − $99 − $129 = money left in budget

   d. $300 − $100 − $100 = $100

   e. $310 − $99 − $129 = $82
   $82 is close to $100, so $82 is a reasonable answer.

2. **a.** What was the total number of donuts?

   b. Needed: the numbers of donuts she bought— 2 dozen (24), 8, and 10

   c. Addition: 24 + 8 + 10 = number of donuts

   d. 20 + 10 + 10 = 40

   e. 24 + 8 + 10 = 42
   42 is close to 40, so 42 donuts is a reasonable answer.

3. **a.** What is each person's share of the rent?

   b. Needed: $386 (the monthly rent) and 2 (the number of people sharing the apartment)
   Not needed: $42 (the amount of the phone bill)

   c. Divide: $386 ÷ 2 = one person's share

   d. $400 ÷ 2 = $200

   e. $386 ÷ 2 = $193
   $193 is close to $200, so $193 is a reasonable answer.

4. **a.** How much will Tim pay for the plywood?

   b. Needed: $26 (cost of a sheet of plywood) and 4 (the number of plywood sheets needed)
   Not needed: $12 (the cost of stain and varnish)

   c. Multiply: $26 × 4 = total cost of plywood

   d. $25 × 4 = $100

   e. $26 × 4 = $104
   $104 is close to $100, so $104 is a reasonable answer.

Using Your Calculator pp. 30–31

Estimates will vary.

1. 972   104   14   3,001
2. 50   44,203   5,625   219
3. 146   810   809   157
4. **685 square yards**
   1,500 − 390 − 425 = 685
5. **$1,500**

   Two approaches are:

   1. Multiply $1,375 by 12 to find the annual salary after the raise ($16,500). Then subtract last year's annual salary. $16,500 − $15,000 = $1,500

   2. Divide $15,000 by 12 to find last year's monthly salary ($1,250). Then subtract it from this year's monthly salary ($1,375) to find the amount of the monthly raise ($125). Multiply this figure by 12. $125 × 12 = $1,500

Unit 1 Review pp. 32–33

Part A

Estimates will vary.

1. 77   797 $1,781   6,462 $28,965
2. 34 $133   324   1,545 $1,801
3. 86   795 $5,480   2,314 $118,350
4. 128   912   108 R2   619   2,809 R5

Part B

Estimates will vary.

5. 595,200   6,540   206

6. 1,806   327   48,888

7. $4,896   9,005   23,865

Part C

8. (3) $176 × 36

   Multiply the amount of each payment by the number of payments.

9. (1) $13 × 25

   Multiply her hourly wage by the number of hours she will work. You don't need the $7 wage.

10. (2) 575 – 492

    Subtract to find the difference between two amounts.

11. (2) 66,411 – 65,847 – 169

    First find the difference between the final odometer reading and the first reading. The difference is the miles she traveled over both days. Subtract the amount she traveled the first day, and the remaining amount is her second day's mileage.

Part D

12. 14,700

    Multiply 2,100 by 7 (the number of sections in the bookstore).

13. 715

    Subtract the number of literature books in stock (1,685) from 2,400.

14. The adventure books were most popular because there are fewer left than any other kind. The reference books were least popular. There are more in stock than any other kind of book.

Working Together

Answers will vary.

Unit 2

Understanding Decimals pp. 36–37

Part A

1. 58.167

2. a. 5
   b. 1
   c. 9
   d. 8

3. a. 0
   b. 6
   c. 6
   d. 5

Part B

4. (2) 4

5. (2) hundredths of a mile

Part C

6. 0.208

   Your explanation may be similar to this:

   The tenths digit must be less than 3 and greater than 1, so put a 2 in the tenths place. Since there are no hundredths, put 0 in the hundredths place. The thousandths place is 4 × 2, so put 8 in the thousandths place. The fourth display reads 0.208.

Writing Decimals pp. 38–39

Part A

1. 1.4

2. 1.04

3. 12.8

4. 12.08

5. 12.008

6. 2.50

7. 2.050

8. 510.07

9. 510.7

10. 510.007

Part B

11. (2) 0.125 inch

12. (1) 5.35 miles

13. (3) 365.26

14. (1) 2.9%

Part C

15–19.

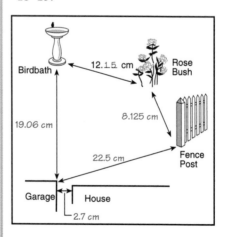

20. Discussion will vary. You may have included some of the following ideas:

The word *point* is useful when numbers must be read aloud. The word *point* is easy to hear and is seldom used incorrectly. Many people use the word *and* incorrectly to read whole numbers. This practice could cause errors.

Decimals and Money pp. 40–41

Part A

1. eighty-five cents

2. three dollars and twenty cents

3. nine dollars and sixty-nine cents

4. fifteen dollars and twenty-seven cents

5. one hundred three dollars and forty-eight cents

6. ninety-five cents

7. seven dollars

8. forty-nine dollars and eight cents

Part B

9. **(3)** One hundred five dollars and nine cents

10. **(1)** $0.89

11. **(1)** $1.35

12. **(2)** $208.53

Part C

13. Deluxe Pizza for $9.95

14. Seven dollars and twenty-five cents

15. Two dollars and seventy-five cents
    $2.75

16. No. He has $8.50. To make $8.60, he would need 6 dimes, not 5.

17. $10.70

Comparing Decimals pp. 42–43

Part A

1. 0.82   0.15   0.411

2. 0.8   0.205   0.5

Part B

3. 0.4 = 0.40   0.43 > 0.054   0.9 > 0.88

4. 0.002 < 0.015   0.9 < 0.95   0.25 > 0.075

Part C

5. Violet

6. Brown

7. Black   White   Brown

8. The wire must have a diameter greater than 0.05 inch and less than 0.1 inch. Only Red (0.06), Green (0.085), Yellow (0.025), and Violet (0.009) are less than 0.1 inch. Of those, Green and Red have diameters greater than 0.05 inch.

Adding Decimals pp. 44–45

Part A

1. 3.75   49.75   46.152   67.38

2. 31.47   19.9   45.01   14.5

Part B

3. $8.97

4. 24.2 ounces

5. 1.763 grams

6. $5.50

Making Connections: Using the Clear Entry Key p. 45

1. 19.669

2. 12.105

3. $171.98

Subtracting Decimals pp. 46–47

Part A

Estimates will vary.

1. 1.9   2.85   5.875   1.156

2. 2.25   7.6   0.718   0.695

3. 0.3   6.98   1.648   20.68

Part B

4. 1.55   2.46   5.8

5. 6.72   1.48   5.357

Part C

6. $6.21

7. $401.68

8. 0.525 second

9. 3.8 degrees

Making Connections: Keeping Mileage Records p. 47

Day	Odometer Reading	Miles Driven
Beginning Reading	34,094.1	
Monday	34,156.9	62.8
Tuesday	34,360.2	203.3
Wednesday	34,615.1	254.9
Thursday	34,704.9	89.8
Friday	34,979.3	274.4
Saturday	35,089.6	110.3
Sunday	35,447.4	357.8

1. Sunday (357.8 miles)
2. You can subtract the beginning odometer reading (34,094.1) from the final reading (35,447.4).

   35,447.4 − 34,094.1 = 1,353.3 miles

Solving Multistep Problems pp. 48–49

1. a. Multistep
   b. Add .277 and .016; then subtract the total from .300.
   c. Answer: .007
2. a. Multistep
   b. Add 5.4 and 3.6; then subtract the total from 13.0.
   c. 4.0
3. a. Single-step
   b. Add $373.90 and $415.75.
   c. $789.65
4. a. Multistep
   b. Add 4.5, 1.25, and 1.75; then subtract 4 from the total.
   c. 3.5 hours

Mixed Review pp. 50–51

Part A

1. (3) four hundred eight thousandths
2. (2) tenths place
3. (1) 3.5 is equal to 3.50.
4. (3) hundredths place
5. (3) One hundred ten dollars and five cents
6. (2) 0.078 is less than 0.780.

Part B

7. 14.3 $160.80 $16.64   3.406
8. 74.75   1.56   51.28   3.375
9. 4.470 *or* 4.47   9.9 $55.52   8.875
10. 1.45   11.23   7.81   23.73

Part C

Estimates will vary.

11. **8.79 inches**

    2.015 + 3.875 + 2.9 = 8.79
12. **0.525 meter**

    1.925 − 1.4 = 0.525
13. **26.2 gallons**

    11.4 + 14.8 = 26.2
14. **$2,125.70**

    $299.95 + $1,825.75 = $2,125.70
15. **Too short; by 0.015 inch**

    0.935 − 0.92 = 0.015
16. **$494.61**

    $204.50 + $489.06 − $198.95 = $494.61

Part D

17. **12.22 miles**

    5.28 + 4.8 + 2.14 = 12.22
18. a. **Olympic to National is the shorter route.**

    Olympic to National: 10.62 + 4.8 = 15.42
    3rd to Crenshaw to National:
    5.925 + 6.375 + 5.28 = 17.58

    b. **This route is 2.16 miles shorter.**

    Find the difference: 17.58 − 15.42 = 2.16 miles
19. **0.48 mile closer to the bank**

    5.28 − 4.8 = 0.48
20. The shortest route is to go on 3rd Street to the park; then take Crenshaw to National to the bank. This route is 22.38 miles long. 5.925 + 6.375 + 5.28 + 4.8 = 22.38 The only other route is to take 3rd Street to the park and back; then take Olympic to the bank. This route is 22.47 miles long. 5.925 + 5.925 + 10.62 = 22.47

Calculators and Decimals pp. 52–53

Part A

1. 20.82   7.125 $2.52
2. 0.263 $17.88 $66.67
3. $4,500 $3.42   60.9

Part B

Estimates will vary.

4. **Multiply, $318.08**

   $8.96 rounds to $9.
   35.5 is close to 40 hours.
   Estimate: $40 \times \$9 = \$360$

5. **Multiply, then add; $21.09**

   15.8 rounds to 16.
   $1.19 rounds to $1.
   $2.29 rounds to $2.
   Estimate: $16 \times \$1 = \$16; \$16 + \$2 = \$18$

6. **Divide, 13 peaches**

   $.36 is close to $.50.
   Estimate: $\$5 \div \$.50 = 10$ peaches

7. **Subtract, then multiply; $33.03**

   $26.99 is close to $25.
   $15.98 is close to $16.
   Estimate: $\$25 - \$16 = \$9$ per shirt; $\$9 \times 3 = \$27$

Making Connections: The Value of a Remainder p. 53

1. **15 client files**

   $775 \div 40 = 19.375$
   Multiply: $19 \times 40 = 760$
   Subtract: $775 - 760 = 15$

2. **$154**

   $\$6,524 \div \$182 = 35.846153$
   Multiply: $35 \times \$182 = \$6,370$
   Subtract: $\$6,524 - \$6,370 = \$154$

Multiplying Decimals pp. 54–55

Part A

1. 78.75   17.0   .0042   .0058   1.14

Part B

Estimates will vary.

2. 11.97   17.86   10.25   44.625

3. 2.8   14.63   0.8125   594

Part C

4. **$19.80**

   $4.5 \times \$4.40 = \$19.80$

5. **$178.25**

   $575 \times \$0.31 = \$178.25$

6. **$4.32**

   $24 \times \$0.18 = \$4.32$

7. **9.25 centimeters**

   $11 \times 8.25 = 90.75$
   $100 - 90.75 = 9.25$

Part D

8. 65   34.7   1,680   5,320

9. 1,005   30.32   1,125   37.5

Dividing Decimals pp. 56–57

Part A

1. 1.06   0.025   8.5   82

2. 0.32   2.35   0.005   19

Part B

3. 1.572   2.058   2.55   0.0006

Part C

4. **6.2 inches**

   $6.2 + 8.5 + 3.9 = 18.6$
   $18.6 \div 3 = 6.2$

5. **8.125 kilometers**

   $8.25 + 6.5 + 7.25 + 10.5 = 32.5$
   $32.5 \div 4 = 8.125$

6. **$38.67**

   $\$154.68 \div 4 = \$38.67$

7. **9 servings**

   $15.75 \div 1.75 = 9$

8. **7.35 hours**

   $7.75 + 6.5 + 7.25 + 8.0 + 7.25 = 36.75$
   $36.75 \div 5 = 7.35$

9. **$9.20 per hour**

   $\$338.10 \div 36.75 = \$9.20$
   You need to divide his weekly paycheck by the number of hours he will work during the week. The answer to the division problem is 9.2. Since the decimal is an amount of money, you need to write a zero in the hundredths place. $\$9.2 = \$9.20$

Choosing a Method pp. 58–59

Part A

1. c

2. b

3. a

4. d

Part B

5. **(3)** 80($39.50) + $252.80

6. **(3)** $\frac{(45.9 + 42.9 + 53.7)}{3}$

7. **(1)** $12.69(20 + 15.8)

8. **(1)** 2(45 + 25)

Making Connections: Finding Another Way p. 59

1. 10(25) − 10(15)

2. 4(8 + 2)

3. 2(16) + 2(9)

Figuring Unit Price and Total Cost pp. 60–61

Part A

1. **$.85**

   $3.40 ÷ 4 = $.85

2. **$.40**

   $2.00 ÷ 5 = $.40

3. **$.055**

   $5.50 ÷ 100 = $.055 This amount could also be read $5\frac{1}{2}$ cents. For very inexpensive items, the unit price may include a fraction of a cent.

4. **$1.29**

   $3.87 ÷ 3 = $1.29

Part B

5. **$10.47**

   $3.49 × 3 = $10.47

6. **$3.36**

   24 × $.14 = $3.36

7. **$27.57**

   $1.49 × 18.5 = $27.565, which rounds to $27.57

8. **$3.45**

   $.69 × 5 = $3.45

Part C

	A & R Unit Price	Foodtime Unit Price	Best Buy
Lucky Cereal	$.06 per oz.	$.07 per oz.	A & R
Boneless Ham	$4.25 per lb.	$4.10 per lb.	Foodtime
Toilet Paper	$.34 per roll	$.28 per roll	Foodtime
Apples	$.50 per lb.	$.55 per lb.	A & R
Baby Shampoo	$.24 per fl. oz.	$.26 per fl. oz.	A & R

9. Answers will vary. You may have mentioned factors such as travel time to store, length of lines at the checkout counter, store's hours, and so on. All of these factors can be measured mathematically and compared.

Unit 2 Review pp. 62–63

Part A

Estimates will vary.

1. 2.35   180.25   2.04

2. 8.5   36.11   8.015

3. 10.143   26   206.04

4. 4   0.42   100

5. 36.8   6.8   28.8

Part B

Estimates will vary.

6. **(2) $250 + 24($39.52)**

   You need to multiply 24 by the amount of the monthly payment and add the down payment.

7. **(2) The unit price for both brands is the same.**

   Brand A: $7.68 ÷ 12 = $.64
   Brand B: $6.40 ÷ 10 = $.64

8. **(1) C, E, A, B, D**

   Work from left to right and compare each digit.

9. **(4) $21.92**

   Divide the total bill by 4. *Note:* One roommate will need to pay $21.93 to cover the bill.

10. **(4) 20**

    $35 ÷ $1.72 = 20.348 Ignore the remainder. You can't buy part of a tube of paint.

11. **(1) 8($1.95) − $5.25**

    You need to multiply the price of a slice by the number of slices to find the total revenue; then subtract the cost to the restaurant for the pie.

Part C

12. **$301.72**

    $267.22 − $15.50 + $50.00 = $301.72

13. **$64.94**

    $56.91 + $68.90 + $72.20 + $61.75 = $259.76
    $259.76 ÷ 4 = $64.94

14. **$118.28**

    Kate has $301.72. Subtract to find the amount she needs to cover the $400 check. $400 − $301.72 = $98.28

    Add $20, the amount she wants to leave in her account. $98.28 + $20 = $118.28

15. **$13,722.80**

    $263.90 × 52 = $13,722.80

Working Together

Answers will vary.

Unit 3

Relating Decimals and Fractions pp. 66–67

Part A

1. 0.4 $\frac{4}{10}$

2. 0.35 $\frac{35}{100}$

3. 0.9 $\frac{9}{10}$

Part B

4. 0.69 0.5 0.482 0.03

5. 0.2 0.25 0.1 0.009

Part C

6. $\frac{50}{100} > 0.48$ $0.6 = \frac{6}{10}$ $0.035 < \frac{350}{1,000}$

7. $\frac{5}{1,000} = 0.005$ $\frac{16}{100} < 0.20$ $0.9 > \frac{8}{10}$

Part D

8. (2) 0.08

9. (1) $\frac{3}{10}$

10. (3) 0.85

11. (3) the swing shift

12. The morning shift had a lower error rate because $\frac{6}{1,000}$ is less than $\frac{6}{100}$. Think about dividing two equal circles into 1,000 and 100 parts. Six of the thousandths would take up much less space than 6 of the hundredths.

13. The morning shift had the lowest defect rate.

Different Forms of Fractions pp. 68–69

Part A

1. $1\frac{2}{5}$ $1\frac{1}{3}$ $5\frac{1}{2}$ 1

2. 2 $1\frac{5}{7}$ $1\frac{1}{4}$ $2\frac{1}{3}$

Part B

3. $\frac{9}{2}$ $\frac{16}{5}$ $\frac{4}{3}$ $\frac{35}{8}$

4. $\frac{8}{3}$ $\frac{8}{5}$ $\frac{23}{6}$ $\frac{23}{4}$

Making Connections: Calculators and Fractions p. 69

1. 0.6

2. 0.875

3. 0.5

4. 0.125

5. 0.75

6. 0.4

Equivalent Fractions pp. 70–71

Part A

1. $\frac{6}{10}$ $\frac{2}{12}$ $\frac{10}{16}$ $\frac{2}{6}$

2. $\frac{3}{12}$ $\frac{9}{24}$ $\frac{27}{30}$ $\frac{9}{21}$

Part B

3. $\frac{4}{7}$ $\frac{2}{5}$ $\frac{2}{8}$ $\frac{6}{12}$

4. $\frac{2}{3}$ $\frac{5}{6}$ $\frac{7}{9}$ $\frac{3}{10}$

Part C

5. $\frac{4}{5}$ $\frac{2}{3}$ $\frac{1}{4}$ $\frac{5}{7}$

6. $\frac{5}{6}$ $\frac{8}{15}$ $\frac{2}{3}$ $\frac{1}{4}$

Part D

7. **#5D**

   $\frac{8}{32}$ inch simplifies to $\frac{1}{4}$ inch.

8. **Just right**

   $\frac{20}{64}$ inch simplifies to $\frac{5}{16}$ inch.

9. $\frac{16}{32}$

   Multiply $\frac{1}{2}$ by $\frac{16}{16}$.
   $\frac{1}{2}$ inch is equal to $\frac{16}{32}$ inch.

English and Metric Rulers pp. 72–73

Part A

1. $1\frac{3}{8}$ in.

2. 5 cm 8 mm or 5.8 cm

3. $4\frac{3}{8}$ in.

4. 11 cm 5 mm or 11.5 cm

Part B

5. Point A $= \frac{7}{8}$ in.   Point B $= 2\frac{1}{4}$ in.
   Point C $= 4\frac{3}{4}$ in.   Point D $= 5\frac{1}{8}$ in.

Part C

6. **The $1\frac{1}{4}$-inch bolt**

   Look at the English ruler. $\frac{1}{4}$ inch is equal to $\frac{4}{16}$ inch. The $1\frac{1}{4}$-inch bolt is slightly longer than the length Jim needs.

7. **The 3.5-cm package**

   Look at the metric ruler. There are 10 millimeters in a centimeter. 35 millimeters is equal to 3.5 centimeters.

Adding and Subtracting Like Fractions pp. 74–75

Part A

1. $\frac{3}{5}$ $\frac{5}{4} = 1\frac{1}{4}$ $\frac{4}{10} = \frac{2}{5}$ $\frac{15}{9} = 1\frac{6}{9} = 1\frac{2}{3}$

2. $\frac{2}{8} = \frac{1}{4}$ $\frac{12}{2} = 6$ $\frac{4}{12} = \frac{1}{3}$ $\frac{10}{6} = 1\frac{4}{6} = 1\frac{2}{3}$

Part B

3. **(3) 1 hour**

$$\frac{3}{4} + \frac{1}{4} = \frac{4}{4} = 1$$

4. **(2) $\frac{1}{2}$ inch**

$$\frac{9}{16} - \frac{1}{16} = \frac{8}{16} = \frac{1}{2}$$

5. **(3) $\frac{1}{20}$ second**

$$\frac{12}{100} - \frac{7}{100} = \frac{5}{100} = \frac{1}{20}$$

6. **(2) $\frac{3}{4}$ cup**

$$\frac{1}{8} + \frac{5}{8} = \frac{6}{8} = \frac{3}{4}$$

Part C

7. **1 hour**

$$\frac{1}{4} + \frac{3}{4} = \frac{4}{4} = 1$$

8. **No.**

$$\frac{5}{4} + \frac{4}{4} = \frac{9}{4} = 2\frac{1}{4} \text{ hours}$$

9. hard drive installation
   sound card installation
   CD-ROM installation
   RAM upgrade
   hard drive and sound card installations
   RAM upgrade and sound card installation
   2 sound card installations
   3 sound card installations

Finding Common Denominators pp. 76–77

Part A

1. $\frac{3}{8} = \frac{6}{16}$

   $\frac{1}{2} = \frac{6}{12}$

   $\frac{2}{7} = \frac{6}{21}$

   $\frac{5}{8} = \frac{25}{40}$

   $\frac{5}{6} = \frac{15}{18}$

2. $\frac{2}{9} = \frac{8}{36}$

   $\frac{3}{5} = \frac{30}{50}$

   $\frac{1}{4} = \frac{12}{48}$

   $\frac{2}{3} = \frac{22}{33}$

   $\frac{7}{10} = \frac{70}{100}$

3. $\frac{8}{11} = \frac{40}{55}$

   $\frac{3}{4} = \frac{15}{20}$

   $\frac{4}{5} = \frac{80}{100}$

   $\frac{4}{7} = \frac{16}{28}$

   $\frac{5}{12} = \frac{25}{60}$

Part B

4. 3: 3, 6, 9, 12, (15)
   5: 5, 10, (15), 20, 25
   2: 2, 4, 6, 8, (10)
   5: 5, (10), 15, 20, 25
   4: 4, 8, (12), 16, 20
   6: 6, (12), 18, 24, 30

5. 4: 4, 8, 12, 16, (20)
   10: 10, (20), 30, 40, 50
   6: 6, 12, (18), 24, 30
   9: 9, (18), 27, 36, 45
   9: 9, 18, 27, (36), 45
   12: 12, 24, (36), 48, 60

Part C

6.

Joe worked longer. Joe worked $\frac{3}{4}$ or $\frac{6}{8}$ hour. Fran worked $\frac{5}{8}$ hour. $\frac{6}{8}$ is greater than $\frac{5}{8}$ so $\frac{3}{4}$ is greater than $\frac{5}{8}$.

Adding and Subtracting Unlike Fractions
pp. 78–79

Part A

1. $\frac{7}{9}$ $\frac{13}{8} = 1\frac{5}{8}$ $\frac{17}{18}$ $\frac{12}{10} = 1\frac{2}{10} = 1\frac{1}{5}$

2. $\frac{1}{12}$ $\frac{1}{18}$ $\frac{9}{20}$ $\frac{1}{15}$

Part B

3. **(2) $\frac{1}{4}$ inch**

$$\frac{5}{16} - \frac{1}{16} = \frac{4}{16} = \frac{1}{4}$$

4. **(2) $\frac{13}{20}$ gallon**

$$\frac{18}{20} - \frac{5}{20} = \frac{13}{20}$$

5. **(3) $\frac{17}{24}$ pound**

$$\frac{9}{24} + \frac{8}{24} = \frac{17}{24}$$

6. **(1) $\frac{1}{4}$ hour**

$$\frac{3}{4} - \frac{2}{4} = \frac{1}{4}$$

Part C

7. $\frac{3}{6} - \frac{3}{8} = \frac{1}{2} - \frac{3}{8} = \frac{4}{8} - \frac{3}{8} = \frac{1}{8}$ **pizza more**

8. 1 or 2 slices of the 8-slice pizza
   1 slice of the 6-slice pizza
   1 slice of the 6-slice pizza and 1 slice of the 8-slice pizza

Working with Distances pp. 80–81

Part A

1. $6\frac{1}{6}$ $3\frac{5}{24}$ $8\frac{1}{10}$ $7\frac{5}{18}$

2. $2\frac{5}{12}$ $2\frac{5}{8}$ $5\frac{19}{30}$ $1\frac{10}{21}$

Part B

3. $1\frac{1}{8} + 1\frac{3}{4} + \frac{3}{4} = 3\frac{5}{8}$ miles

4. $1\frac{1}{4} + 1\frac{3}{8} = 2\frac{5}{8}$
$1\frac{1}{8} + 2\frac{1}{4} = 3\frac{3}{8}$
$3\frac{3}{8} - 2\frac{5}{8} = \frac{6}{8} = \frac{3}{4}$ mile

5. $1\frac{1}{4} + 1\frac{3}{4} = 3$
$\frac{1}{2} + \frac{3}{4} = 1\frac{1}{4}$
$3 - 1\frac{1}{4} = 1\frac{3}{4}$ miles

6. Mall to grocery store to home to park is shorter.
Mall to grocery store to home to park $=$
$1\frac{3}{8} + \frac{1}{2} + \frac{3}{4} = 1\frac{13}{8} = 2\frac{5}{8}$ miles
mall to office to park $= 1\frac{7}{8} + 1\frac{1}{2} = 3\frac{3}{8}$ miles

Making Connections: Distance and Exercise p. 81

1. Home to jewelry store to bakery to bus station to home $= 2$ miles *or*
home to jewelry store to bakery to movie theater to bus station to home $= 2$ miles (Or he could follow these paths in the opposite direction.)

2. Answers will vary. Total miles should be $2\frac{1}{2}\left(\frac{20}{8}\right)$ or more. Sample answers are provided.
Home to jewelry store to bakery to movie theater to bus station to bakery to jewelry store to home $=$
$2\frac{6}{8} = 2\frac{3}{4}$ miles $\left(\frac{22}{8}\right)$ *or*
home to bus station to bakery to movie theater to bus station to home $= 2\frac{3}{4}$ miles $\left(\frac{22}{8}\right)$

Multiplying Fractions pp. 82–83

Part A

1. $\frac{1}{5}$ $\frac{7}{12}$ $\frac{3}{5}$ $\frac{5}{48}$

2. $\frac{1}{4}$ $\frac{3}{5}$ $\frac{1}{6}$ $\frac{2}{7}$

Part B

3. (3) $\frac{1}{3}$ yard
$\frac{2}{3} \times \frac{1}{2} = \frac{1}{3}$

4. (1) $\frac{7}{12}$ pound
$\frac{7}{8} \times \frac{2}{3} = \frac{7}{12}$

5. (3) $\frac{3}{8}$ cup
$\frac{3}{4} \times \frac{1}{2} = \frac{3}{8}$

6. (3) $\frac{3}{16}$ fluid ounce
$\frac{15}{16} \times \frac{1}{5} = \frac{3}{16}$

Part C

7. $\frac{1}{6}$
$\frac{1}{2} \times \frac{1}{3} = \frac{1}{6}$

8. 4 students
$\frac{1}{6} \times 24 = 4$

9. a. $\frac{1}{4}$; 1 student per smaller group ($\frac{1}{4} \times 4 = 1$) or 3 students total

b. $\frac{1}{2}$; 6 students (larger group: $\frac{1}{2} \times 24 = 12$; $\frac{1}{2} \times 12 = 6$)

10. Discussion will vary. You should find that multiplying an amount by a fraction means finding a fraction of that amount.

Dividing Fractions pp. 84–85

Part A

1. $\frac{15}{16}$ $\frac{5}{4} = 1\frac{1}{4}$ $\frac{2}{3}$ $\frac{4}{3} = 1\frac{1}{3}$

2. $\frac{1}{12}$ $\frac{1}{18}$ 18 $\frac{14}{15}$

Part B

3. larger smaller smaller larger

Making Connections: Estimating with Fractions p. 85

1. estimate: $\frac{1}{2} \times 1 = \frac{1}{2}$
exact: $\frac{18}{35}$

2. estimate: $1 \times \frac{1}{2} = \frac{1}{2}$
exact: $\frac{7}{24}$

3. estimate: $1 \div \frac{1}{2} = 2$
exact: $\frac{28}{15} = 1\frac{13}{15}$

4. estimate: $1 \div 9 = \frac{1}{9}$
exact: $\frac{1}{12}$

Dividing Fractions with Mixed Numbers
pp. 86–87

Part A

1. 2 $\frac{17}{2} = 8\frac{1}{2}$ $\frac{184}{25} = 7\frac{9}{25}$ $\frac{8}{21}$

2. $\frac{1}{4}$ $\frac{3}{2} = 1\frac{1}{2}$ $\frac{8}{3} = 2\frac{2}{3}$ $\frac{7}{8}$

3. $\frac{1}{2}$ $\frac{3}{64}$ $\frac{3}{2} = 1\frac{1}{2}$ $\frac{20}{3} = 6\frac{2}{3}$

4. $\frac{22}{25}$ $\frac{15}{2} = 7\frac{1}{2}$ $\frac{37}{12} = 3\frac{1}{12}$ 4

Part B

5. (2) $3\frac{1}{16}$ glasses
$12\frac{1}{4} \div 4 = 3\frac{1}{16}$

6. (1) 32 servings
$10\frac{2}{3} \div \frac{1}{3} = \frac{32}{3} \times \frac{3}{1} = 32$

7. (3) 16 presents
$24 \div 1\frac{1}{2} = 24 \times \frac{2}{3} = 16$

8. (2) 13 years
$6\frac{1}{2} \div \frac{1}{2} = \frac{13}{2} \times \frac{2}{1} = \frac{13}{1} = 13$

Making Connections: Estimating and Mixed Numbers p. 87

Estimates will vary.

1. estimate: $4\frac{1}{3} \div 2 = \frac{13}{6} = 2\frac{1}{6}$
   exact: $\frac{104}{51} = 2\frac{2}{51}$

2. estimate: $6 \div 1 = 6$
   exact: 7

3. estimate: $4 \div 2 = 2$
   exact: $\frac{30}{13} = 2\frac{4}{13}$

4. estimate: $5\frac{1}{2} \div 1 = 5\frac{1}{2}$
   exact: $\frac{22}{3} = 7\frac{1}{3}$

Mixed Review pp. 88–89

Part A

1. (2) $\frac{4}{3}$
2. (3) $\frac{3}{4}$
3. (3) 5.8 centimeters
4. (1) $\frac{9}{12}$
5. (2) fifteenths
6. (1) $1 \times \frac{1}{2}$

Part B

7. $\frac{8}{12} = \frac{2}{3}$ $\frac{6}{4} = 1\frac{1}{2}$ $\frac{9}{8} = 1\frac{1}{8}$ $\frac{19}{18} = 1\frac{1}{18}$
8. $\frac{6}{8} = \frac{3}{4}$ $\frac{2}{3}$ $\frac{10}{12} - \frac{9}{12} = \frac{1}{12}$ $2\frac{27}{18} - \frac{14}{18} = 2\frac{13}{18}$
9. $\frac{1}{9}$ $\frac{1}{2}$ $\frac{7}{15}$ $\frac{6}{35}$
10. $\frac{2}{3}$ $\frac{35}{18} = 1\frac{17}{18}$ $\frac{3}{8}$ 12

Part C

11. $3\frac{1}{3} + 2\frac{1}{2} = 5\frac{5}{6}$ **pounds**
12. $60\frac{1}{2} - 56\frac{3}{4} = 3\frac{3}{4}$ **inches**
13. Estimates may vary.
   $34\frac{1}{4} \times \frac{3}{4}$
   $34 \times \frac{3}{4} = 25\frac{1}{2}$ The $34\frac{1}{4}$-fluid-ounce bottle still has more juice in it.
14. $\frac{16}{3} \times \frac{1}{4} = \frac{4}{3} = 1\frac{1}{3}$ **pounds**
15. $\frac{2}{3} \times \frac{1}{6} = \frac{1}{9}$ **of the whole pie**
16. $\frac{13}{4} \times \frac{3}{1} = \frac{39}{4} = 9\frac{3}{4}$ She can pot **9 plants** (with enough soil for $\frac{3}{4}$ of a pot left over).

Part D

17. $2\frac{2}{8} + 1\frac{4}{8} + \frac{3}{8} + 3\frac{2}{8} = 6\frac{11}{8} = 7\frac{3}{8}$ **miles**
18. $2\frac{1}{4} - 1\frac{2}{4} = \frac{3}{4}$ **mile**
19. $\frac{13}{4} \times \frac{2}{3} = \frac{13}{6} = 2\frac{1}{6}$ **miles**

20. The length of the entire loop not counting the beauty salon is $2\frac{1}{4} + 1\frac{1}{4} + 3\frac{1}{4} = 6\frac{3}{4}$. Subtract this distance from the distance found in problem 17 to get $\frac{5}{8}$. You can also subtract $1\frac{1}{4}$ from the distance from the restaurant to the municipal building via the beauty salon $\left(1\frac{1}{2} + \frac{3}{8} = 1\frac{7}{8}\right)$ to get $\frac{5}{8}$. Either way, $\frac{5}{8}$ **mile** is saved by not going to the beauty salon.

Relating Fractions and Ratios pp. 90–91

Part A

1. $\frac{3}{6} = \frac{1}{2}$ or 1:2 or 1 to 2
2. $\frac{4}{1}$ or 4:1 or 4 to 1
3. $\frac{3}{4}$ or 3:4 or 3 to 4
4. $\frac{1}{6}$ or 1:6 or 1 to 6
5. $\frac{6}{4} = \frac{3}{2}$ or 3:2 or 3 to 2
6. $\frac{4}{6} = \frac{2}{3}$ or 2:3 or 2 to 3

Part B

7. $\frac{\text{circles}}{\text{wavy lines}} = \frac{6}{5}$

8–12. Variety of answers are possible.

Part C

13. $\frac{11}{30}$ or 11:30 or 11 to 30
14. Any ratios involving Hispanic students or the total number of students will be changed.
15. $\frac{8}{5}$ or 8:5 or 8 to 5
16. Drawings will vary but should represent the ratio 6 to 8 (or 3 to 4).

Writing Ratios pp. 92–93

Part A

1. $\frac{20}{18} = \frac{10}{9}$ or 10:9 or 10 to 9
2. $\frac{20}{36} = \frac{5}{9}$ or 5:9 or 5 to 9
3. $\frac{36}{18} = \frac{2}{1}$ or 2:1 or 2 to 1

Part B

4. $\frac{8}{16} = \frac{1}{2}$ or 1:2 or 1 to 2
5. $\frac{2}{2} = \frac{1}{1}$ or 1:1 or 1 to 1
6. $\frac{12}{16} = \frac{3}{4}$ or 3:4 or 3 to 4
7. $\frac{8}{4} = \frac{2}{1}$ or 2:1 or 2 to 1
8. $\frac{2}{16} = \frac{1}{8}$ or 1:8 or 1 to 8

Part C

9. $\frac{3}{6} = \frac{1}{2}$ or 1:2 or 1 to 2
10. $\frac{9}{45} = \frac{1}{5}$ or 1:5 or 1 to 5
11. $\frac{10}{15} = \frac{2}{3}$ or 2:3 or 2 to 3

Writing Proportions pp. 94–95

Part A

1. yes; $24 = 24$
   no; $8 \neq 6$
   yes; $24 = 24$
   yes; $9 = 9$

2. yes; $56 = 56$
   no; $72 \neq 36$
   no; $12 \neq 16$
   yes; $30 = 30$

3. yes; $60 = 60$
   no; $24 \neq 18$
   yes; $30 = 30$
   yes; $216 = 216$

Part B

4. $3 \times a = 24$
   $a = \mathbf{8}$
   $3 \times b = 18$
   $b = \mathbf{6}$
   $16 \times c = 80$
   $c = \mathbf{5}$
   $15 \times d = 30$
   $d = \mathbf{2}$

5. $5 \times e = 105$
   $e = \mathbf{21}$
   $8 \times f = 72$
   $f = \mathbf{9}$
   $8 \times g = 16$
   $g = \mathbf{2}$
   $6 \times h = 12$
   $h = \mathbf{2}$

6. $16 \times i = 16$
   $i = \mathbf{1}$
   $10 \times j = 300$
   $j = \mathbf{30}$
   $25 \times k = 25$
   $k = \mathbf{1}$
   $21 \times l = 126$
   $l = \mathbf{6}$

7. $7 \times m = 392$
   $m = \mathbf{56}$
   $5 \times n = 150$
   $n = \mathbf{30}$
   $20 \times p = 120$
   $p = \mathbf{6}$
   $96 \times q = 480$
   $q = \mathbf{5}$

Making Connections: Making a Table p. 95

1.

Cups	4	16
Muffins	20	x

$\frac{4}{20} = \frac{16}{x}$
$4 \times x = 320$
$x = \mathbf{80}$

2.

Home runs	5	x
Games	12	60

$\frac{5}{12} = \frac{x}{60}$
$12 \times x = 300$
$x = \mathbf{25}$

Solving Problems with Proportions pp. 96–97

Part A

1. (2) and (4)

2. (3) and (4)

3. (1) and (4)

4. (1) and (3)

Part B

5. $\frac{1.5}{1} = \frac{6}{x}$
   $1.5 \times x = 6$
   $x = \mathbf{4 \text{ tablespoons}}$

6. $\frac{8}{24} = \frac{x}{222}$
   $\frac{1}{3} = \frac{x}{222}$
   $3x = 222$
   $x = \mathbf{74 \text{ blue tiles}}$

Part C

7. $\frac{4}{220} = \frac{5}{x}$
   $\frac{1}{55} = \frac{5}{x}$
   $x = \mathbf{275 \text{ rushing yards}}$

8. $\frac{5}{1,500} = \frac{4}{x}$
   $\frac{1}{300} = \frac{4}{x}$
   $x = \mathbf{1,200 \text{ passing yards}}$

9. Ratios and proportions can be used to predict an outcome, but they do not determine it. In your answer, you may have mentioned various nonstatistical or nonmathematical factors, such as emotion, weather, health (injuries), etc.

Understanding Percents pp. 98–99

Part A

1. 70%   33%   96%   50%

2. 11%   70%   84%   80%

Part B

3. 65%

4. 35%

5. 75%

6. 45%

Part C

7. 16%

8. 78%

9. 31%

10. 47%

11. Answers will vary. You may have listed such things as taxes, tips, discounts, sports, advertisements, survey results, weather reports, food packages, test scores, elections, and interest on loans or savings accounts.

Decimals, Fractions, and Percents pp. 100–101

Part A

1. .16   3   .5   .978

2. .03   .082   .007   12.59

Part B

3. 67%   1%   40%   340%

4. 506%   419%   8.2%   2,380%

Part C

5. $\frac{75}{100} = \frac{3}{4}$ $\frac{10}{100} = \frac{1}{10}$ $\frac{50}{100} = \frac{1}{2}$ $\frac{12}{100} = \frac{3}{25}$

6. 75%   37.5%   60%   66.7%

Part D

7. 62.5%

8. 20%

9. $\frac{60}{100} = \frac{3}{5}$

10.

Fraction	Percent	Decimal
$\frac{1}{2}$	50%	.5
$\frac{1}{3}$	$33\frac{1}{3}$ %	$.3333\ldots = .33\frac{1}{3}$
$\frac{2}{3}$	$66\frac{2}{3}$ %	$.6666\ldots = .66\frac{2}{3}$
$\frac{1}{4}$	25%	.25
$\frac{3}{4}$	75%	.75
$\frac{1}{5}$	20%	.2
$\frac{2}{5}$	40%	.4
$\frac{3}{5}$	60%	.6
$\frac{4}{5}$	80%	.8
$\frac{1}{10}$	10%	.1
$\frac{3}{10}$	30%	.3
$\frac{7}{10}$	70%	.7
$\frac{9}{10}$	90%	.9

The Percent Equation pp. 102–103

Part A

1. $.45 \times 62 = p$
   $.23 \times 134 = p$
   $.89 \times 1{,}530 = p$
   $.09 \times 431 = p$

2. $.75 \times w = 65$
   $.06 \times w = 128$
   $1 \times w = 40$
   $.18 \times w = 850$

3. $n\% \times 35 = 20$
   $n\% \times 174 = 9$
   $n\% \times 521 = 502$
   $n\% \times 67 = 13$

Part B

4. 38% of $1,500 is ___.
   $.38 \times 1{,}500 = p$

5. 65% of ___ is 262.
   $.65 \times w = 262$

6. ___ of 87 is 34.
   $n\% \times 87 = 34$

7. 29% of 734 is ___.
   $.29 \times 734 = p$

Making Connections: Circle Graphs p. 103

1. $100\% - 91\% = \mathbf{9\%}$

2. 48% of 481 is ___.
   $.48 \times 481 = p$

Solving Percent Equations pp. 104–105

Part A

1. $.8 \times w = 1{,}200$
   $w = \mathbf{1{,}500}$
   $n\% \times 55 = 11$
   $n\% = .2 = \mathbf{20\%}$
   $.35 \times 60 = p$
   $p = \mathbf{21}$
   $.15 \times w = 6$
   $w = \mathbf{40}$

2. $n\% \times 18 = 9$
   $n\% = .5 = \mathbf{50\%}$
   $.9 \times 80 = p$
   $p = \mathbf{72}$
   $.24 \times w = 1{,}647$
   $w = \mathbf{6{,}862.5}$
   $n\% \times 748 = 561$
   $n\% = .75 = \mathbf{75\%}$

3. $.62 \times 114 = p$

$p = \textbf{70.68}$

$.25 \times w = 2{,}498$

$w = \textbf{9,992}$

$n\% \times 180 = 120$

$n\% = .66\frac{2}{3} = \textbf{66}\frac{\textbf{2}}{\textbf{3}}\textbf{\%}$

$.07 \times 584 = p$

$p = \textbf{40.88}$

Part B

4. $.17 \times w = 43$

$w = \textbf{\$252.94}$

5. $.15 \times 839 = p$

$p = \textbf{\$125.85}$

6. $n\% \times 1{,}165 = 128$

$n\% = \textbf{11\%}$

7. $.9 \times w = 1{,}350$

$w = \textbf{1,500}$

Part C

8. $.32 \times \$1{,}900 = p$

$p = \textbf{\$608}$

9. $n\% \times \$1{,}900 = \$627$

$n\% = \textbf{33\%}$

10. $.24 \times \$1{,}900 = p$

$p = \$456$

$.25 \times \$1{,}900 = p$

$p = \$475$

$\$475 - \$456 = \textbf{\$19}$

or

$25\% - 24\% = 1\%$

$.01 \times \$1{,}900 = \textbf{\$19}$

11. Circle graph drawings will vary greatly.

Discounts pp. 106–107

Part A

1. Savings: $3.00

Discount price: $34.45

2. Savings: $97.00

Discount price: $291.00

3. Savings: $1.95

Discount price: $11.03

4. Savings: $84.32

Discount price: $477.83

Part B

5. $351.12

6. $32.42

Making Connections: Using a Calculator to Find Discounts p. 107

Answers should be exactly the same as those shown for problems 5 and 6.

Two-Step Percent Problems pp. 108–109

Part A

1. $112\% \times 10.4 = 11.648$ yards per catch

$11.648 \times 8 = 93.184 \approx \textbf{93 yards}$

You could also use this approach:

$112\% \times 10.4 = 11.648 \approx 12$ per catch

$12 \times 8 = \textbf{96 yards}$

2. $.12 \times w = 240$

$w = \$2{,}000$

$\$2{,}000 \div 20 = \textbf{\$100 per day}$

3. $275 - 230 = 45$ more cars

$45 \div 230 \approx \textbf{20\%}$

4. $.1 \times w = \$14{,}000$

$w = \$140{,}000$

$\$140{,}000 - \$14{,}000 = \textbf{\$126,000}$

Part B

5. $32.19

6. $17.53

7. $133.52

8. $103.85

9. $\$55.85 - 10\% \times \$55.85 =$

$\$55.85 - \$5.59 = \$50.26$

$\$50.26 + 8.25\% \times \$50.26 =$

$\$50.26 + \$4.15 = \textbf{\$54.41}$

or

$100\% - 10\% = 90\%$ paid

$.90 \times \$55.85 = \$50.27$

$\$50.27 + 8.25\% \times \$50.27 =$

$\$50.27 + \$4.15 = \textbf{\$54.42}$

Unit 3 Review pp. 110–111

Part A

1. $\frac{6}{9} = \frac{2}{3}$ $\frac{25}{24} = 1\frac{1}{24}$ $\frac{4}{10} = \frac{2}{5}$ $\frac{11}{24}$

2. $1\frac{11}{12}$ $\frac{8}{15}$ $\frac{1}{2}$ 1

3. $\frac{1}{4}$ $\frac{9}{8} = 1\frac{1}{8}$ 22 $\frac{5}{16}$

Part B

4. **(3)** D, E, C, B, A

5. **(1)** $\frac{1}{6}$

6. **(4)** 4:3

7. **(4)** $.05 \times w = 5.25$

Part C

8. $.75

9. $40 − 15%($40) = $34

   Brian will save $1.00 by buying on sale at Federal.

10. $33\frac{1}{3}\%$

11. $14.88

12. $6.09

13. $23.06 before tax; **$27.13** original price

Part D

14. 145

15. 157

16. 332

Working Together

Circle graph drawings will vary greatly.

Unit 4

English Units of Length pp. 114–115

Estimates will vary.

Part A

1. 28 in. ≈ 2 ft.
   28 in. = 2 ft. 4 in.

   47 in. ≈ 4 ft.
   47 in. = 3 ft. 11 in.

   16 ft. ≈ 5 yd.
   16 ft. = 5 yd. 1 ft.

2. 23 ft. ≈ 8 yd.
   23 ft. = 7 yd. 2 ft.

   81 in. ≈ 2 yd.
   81 in. = 2 yd. 9 in.

   110 in. ≈ 3 yd.
   110 in. = 3 yd. 2 in.

Part B

3. 3 ft. ≈ 36 in.
   3 ft. = 36 in.

   $5\frac{1}{6}$ ft. ≈ 60 in.
   $5\frac{1}{6}$ ft. = 62 in.

   $1\frac{1}{3}$ yd. ≈ 3 ft.
   $1\frac{1}{3}$ yd. = 4 ft.

4. $4\frac{3}{4}$ yd. ≈ 180 in.
   $4\frac{3}{4}$ yd. = 171 in.

   $1\frac{1}{5}$ mi. ≈ 1,760 yd.
   $1\frac{1}{5}$ mi. = 2,112 yd.

   $2\frac{5}{6}$ mi. ≈ 15,840 ft.
   $2\frac{5}{6}$ mi. = 14,960 ft.

Part C

5. 43

6. $6\frac{17}{36}$, rounded to **7 yd.**

7. 78 in.

8. 12,534 × 3 = **37,602 ft.**
   12,534 × 36 = **451,224 in.**
   12,534 ÷ 1,760 ≈ **7 mi.**

Working with Length pp. 116–117

Part A

1. 9 ft. 9 in.
   7 ft. 16 in. = 8 ft. 4 in.
   13 ft. 21 in. = 14 ft. 9 in.
   6 yd. 4 ft. = 7 yd. 1 ft.

2. 3 ft. 4 in.
   2 ft. 10 in.
   4 ft. 3 in.
   4 yd. 2 ft.

Part B

3. 6 ft. 12 in. = 7 ft.
   25 ft. 10 in.
   16 ft. 36 in. = 19 ft.
   42 ft. 60 in. = 47 ft.

4. 2 ft. 4 in.
   1 ft. 4 in.
   2 ft. 5 in.
   1 yd. 2 ft.

Part C

5. 6 ft. 9 in. ÷ 3 = **2 ft. 3 in.**

   With one bush at each head of the flower bed, the remaining bushes separate the bed into 3 equal sections.

6. 4 ft. 2 in. + 5 ft. 4 in. + 5 ft. 9 in. =
   14 ft. 15 in. = 15 ft. 3 in.
   15 ft. 3 in. ÷ 3 = **5 ft. 1 in.**

7. 6 yd. 1 ft. × 3 = 18 yd. 3 ft. = **19 yd.**

8. Whole numbers that divide 3 ft. 8 in. evenly are **2, 4, 11, 22,** and **44.**
   3 ft. 8 in. ÷ 2 = 1 ft. 10 in.
   3 ft. 8 in. ÷ 4 = 11 in.
   3 ft. 8 in. ÷ 11 = 4 in.
   3 ft. 8 in. ÷ 22 = 2 in.
   3 ft. 8 in. ÷ 44 = 1 in.

   Whole numbers that give an even number of feet when multiplying 3 ft. 8 in. are **3, 6, 9, 12,** and **27.**
   3 ft. 8 in. × 3 = 11 ft.
   3 ft. 8 in. × 6 = 22 ft.
   3 ft. 8 in. × 9 = 33 ft.
   3 ft. 8 in. × 12 = 44 ft.
   3 ft. 8 in. × 27 = 99 ft.

Measuring Capacity pp. 118–119

Part A

Estimates will vary.

1. 24 fl. oz. ≈ 3 c.
   24 fl. oz. = 3 c.

   5 c. ≈ 2 pt. *or* 3 pt.
   5 c. = 2 pt. 1 c.

   7 pt. ≈ 3 qt. *or* 4 qt.
   7 pt. = 3 qt. 1 pt.

   14 qt. ≈ 3 gal. *or* 4 gal.
   14 qt. = 3 gal. 2 qt.

2. $2\frac{1}{8}$ c. ≈ 16 fl. oz.
   $2\frac{1}{8}$ c. = 17 fl. oz.

   10 pt. ≈ 20 c.
   10 pt. = 20 c.

   $3\frac{3}{4}$ gal. ≈ 16 qt.
   $3\frac{3}{4}$ gal. = 15 qt.

   $5\frac{1}{2}$ gal. ≈ 24 c.
   $5\frac{1}{2}$ gal. = 22 c.

Part B

3. 5 c. 9 fl. oz. = 6 c. 1 fl. oz.
   4 pt. 2 c. = 5 pt.
   4 qt. 2 pt. = 5 qt.
   7 gal. 5 qt. = 8 gal. 1 qt.

4. 3 c. 3 fl. oz.   3 c. 5 fl. oz.   2 pt. 1 c.   6 gal. 2 qt.

Part C

5. 3 qt. 1 pt. − 2 qt. 1 pt. = **1 qt.**

6. 4 pouches × 6 fl. oz = 24 fl. oz.
   4 pouches × 8 fl. oz. = 32 fl. oz.
   Dave needs to add 24–32 fl. oz. of hot water.
   3 c. × 8 = 24 fl. oz.
   Dave has added enough water.

7. 1 gal. $\times \frac{4 \text{ qt.}}{\text{gal.}} \times \frac{2 \text{ pt.}}{\text{qt.}} \times \frac{2 \text{ c.}}{\text{pt.}}$ = 16 c.
   16 c. − 3 c. = 13 c. = **3 qt. 1 c.**

8.

	Fluid Ounces	Cups
Pint	16	2
Quart	32	4
Gallon	128	16

Using Rulers, Cups, and Spoons pp. 120–121

Part A

1. A = $1\frac{3}{4}$ in.
   B = $2\frac{5}{8}$ in.
   C = $4\frac{1}{2}$ in.
   D = $5\frac{7}{16}$ in.

Part B

2. $2\frac{5}{8} - 1\frac{3}{4} = \frac{7}{8}$ **in.**
   $4\frac{1}{2} - 2\frac{5}{8} = 1\frac{7}{8}$ **in.**
   $5\frac{7}{16} - 2\frac{5}{8} = 2\frac{13}{16}$ **in.**
   $5\frac{7}{16} - 4\frac{1}{2} = \frac{15}{16}$ **in.**

Part C

3. $\frac{5}{8}$ c.
   5 fl. oz.

4. $\frac{4}{8} = \frac{1}{2}$ c.
   4 fl. oz.

Part D

5. $\frac{1}{4}$ c. = 2 fl. oz. $\times \frac{2 \text{ tbsp.}}{\text{fl. oz.}}$ = **4 tbsp.**

6. **Corrine will not have any milk left over.**
   3 tsp. = 1 tbsp. used in coffee
   1 tbsp. + 5 tbsp. = 6 tbsp. =
   3 fl. oz. used altogether
   3 fl. oz. − 3 fl. oz. = 0 (nothing left)

Measuring Weight pp. 122–123

Part A

Estimates will vary.

1. 32 oz. ≈ 2 lb.
   32 oz. = 2 lb.

   40 oz. ≈ 2 lb. *or* 3 lb.
   40 oz. = 2 lb. 8 oz.

   92 oz. ≈ 6 lb.
   92 oz. = 5 lb. 12 oz.

   4,500 lb. ≈ 2 t.
   4,500 lb. = 2 t. 500 lb.

2. $1\frac{1}{2}$ lb. ≈ 32 oz.
   $1\frac{1}{2}$ lb. = 24 oz.

   $2\frac{1}{4}$ lb. ≈ 32 oz.
   $2\frac{1}{4}$ lb. = 36 oz.

   3 t. ≈ 6,000 lb.
   3 t. = 6,000 lb.

   $4\frac{3}{4}$ t. ≈ 10,000 lb.
   $4\frac{3}{4}$ t. = 9,500 lb.

Part B

3. 3 lb. 11 oz.
  10 lb. 25 oz. = 11 lb. 9 oz.
  3 t. 728 lb.
  8 t. 2,750 lb. = 9 t. 750 lb.

4. 2 lb. 4 oz.
  1 lb. 7 oz.
  5 lb. 15 oz.
  2 t. 1,600 lb.

Part C

5. 5 lb. − 2 lb. 6 oz. = **2 lb. 10 oz.**

6. $1\frac{1}{2}$ t. = 3,000 lb.
  3,000 lb. − 1,756 lb. = **1,244 lb.**

7. 35 lb. 3 oz. + 14 lb. 15 oz. =
  49 lb. 18 oz. = **50 lb. 2 oz.**

8. 220 lb. × 9 = 1,980 lb.
  1,980 lb. is less than 1 t. (2,000 lb.). It is safe to load 9 crates onto the freight elevator.

9. Discussion will vary.
  truck—tons and pounds
  man—pounds
  box of cereal—ounces

Using Metric Units pp. 124–125

Part A

1. length: miles and kilometers
  width: feet or yards and meters

2. inches and millimeters or centimeters

3. feet or yards and meters

Part B

4. 194 mm   **7.** 29 mm

5. 870 cm   **8.** 500 m

6. 20,000 m   **9.** 64 cm

Part C

10. pounds and kilograms

11. fluid ounces, cups, and milliliters

12. fluid ounces or quarts and liters

Part D

13. .5 g   **16.** 4.5 kg

14. 1.2 liters   **17.** .8 kg

15. .022 liter   **18.** .025 g

Measuring Temperature pp. 126–127

Part A

Answers will vary.

1. −10°F   −23°C

2. 50°F   10°C

3. 32°F   0°C

4. 86°F   30°C

5. 68°F   20°C

6. 100°F   38°C

Part B

7. 90°F   32°C

8. 20°F   −7°C

9. 60°F   16°C

Part C

10. 67 − 8 = **59°F**

11. 78°F = 25°C
  90°F = 32°C
  32 − 25 = **7°C**

12. 10°C = 50°F
  90 − 50 = **40°F**

13. 120 − (−40) = **160°F**
  50 − (−40) = **90°C**

Making Connections: The Clinical Thermometer p. 127

normal − under the arm = 98.6° − 97° = 1.6° difference

99° + 1.6° = 100.6°F

Your child's temperature is above normal.

Reading Scales and Meters pp. 128–129

Part A

1. A = 1 lb. 8 oz.
    $1\frac{1}{2}$ lb.
  B = 2 lb. 4 oz.
    $2\frac{1}{4}$ lb.
  C = 2 lb. 12 oz.
    $2\frac{3}{4}$ lb.

2. D = 3 lb. 12 oz.
    $3\frac{3}{4}$ lb.
  E = 4 lb. 8 oz.
    $4\frac{1}{2}$ lb.
  F = 5 lb. 4 oz.
    $5\frac{1}{4}$ lb.

3. G = 1 lb. 7 oz.
  H = 3 lb. 2 oz.
  I = 4 lb. 13 oz.

Part B

4.

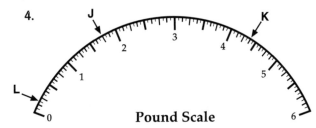

Pound Scale

Part C

5. A = 20 MPH
       32 KPH
   B = 40 MPH
       64 KPH
   C = 60 MPH
       96 KPH

6. D = 13 MPH
       20 KPH
   E = 45 MPH
       70 KPH
   F = 58 MPH
       90 KPH

Part D

7. 32,941.8 miles

8. Discussion will vary. It may include maintaining speed limit, knowing when to change oil, safe driving speeds in bad weather, measuring distance, when to shift gears (manual transmission), and so on.

Figuring Distance, Rate, and Time pp. 130–131

Part A

1. 180 miles   90 miles   96 miles   256 miles

Part B

2. 40 MPH
   43 MPH
   58 MPH
   50 MPH

Part C

3. 4 hours   6 hours   9 hours   11 hours

Part D

4. 48×6 = **288 miles**

5. $1,575 \div 36$ = 43.75, rounded to **44 miles per hour**

6. $30 \div 32$ = .9375 hour, rounded to **1 hour**

7. at 25 miles per hour = 1.8 hours
   at 30 miles per hour = 1.5 hours
   at 35 miles per hour = 1.3 hours
   at 40 miles per hour = 1.1 hours
   at 45 miles per hour = 1 hour

Mixed Review pp. 132–133

Part A

1. 42 in.   2 ft. 10 in.   15 ft.

2. 5 yd. 2 ft.   6,600 ft.   1 c. 4 fl. oz.

3. 22 fl. oz.   2 lb. 14 oz.   4,800 lb.

4. 144 fl. oz.   1 lb. 2 oz.   56 oz.

5. 4,200 m   17,100 mg   2.75 liters

Part B

6. 7 ft. 13 in. = 8 ft. 1 in.
   3 ft. 3 in.
   5 yd. 3 ft. = 6 yd.

7. 5 c. 9 fl. oz. = 6 c. 1 fl. oz.
   2 fl. oz.
   3 gal. 3 qt.

8. 6 gal. 5 qt. = 7 gal. 1 qt.
   7 lb. 20 oz. = 8 lb. 4 oz.
   15 oz.

Part C

9. $1\frac{1}{2}$ c. in the first jar
   10 fl. oz. in the second jar
   $1\frac{1}{2}$ c. = 12 fl. oz.
   12 fl. oz. − 10 fl. oz. = **2 fl. oz. more in the first jar**
   12 fl. oz. + 10 fl. oz. = 22 fl. oz. =
   2 c. 6 fl. oz. combined

10. 3 ft. $1\frac{1}{2}$ in. long now
    3 ft. $1\frac{1}{2}$ in. = 2 ft. $13\frac{4}{8}$ in. = 2 ft. $12\frac{12}{8}$ in.
    2 ft. $12\frac{12}{8}$ in. − $3\frac{7}{8}$ in. = **2 ft. $9\frac{5}{8}$ in. long after it is cut**

Part D

11. $6\frac{1}{4} \times 12$ = **75 in.**

12. 2 t. 500 lb. + 800 lb. = **2 t. 1,300 lb.**

13. 32 oz. = 2 lb. **Claudio has exactly enough meat.**

14. 5 yd. − 2 yd. 2 ft. = **2 yd. 1 ft.**

15. $t = \frac{d}{r}$
    $360 \div 35$ = **10 hours** (rounded)

16. $r = \frac{d}{t}$
    $55 \div 1$ = **55 miles per hour**

17. 1 c. 3 fl. oz. + 2 c. 6 fl. oz. = 3 c. 9 fl. oz. =
    4 c. 1 fl. oz.

18. 1 t. 200 lb. − 750 lb. = **1,450 lb.**

19. 440×8 = 3,520 yd.
    $3,520 \div 1,760$ = **2 mi.**

20. $7,208 \times 3 = $ **21,624 ft.**

21. $55 \times 4 = $ **220 km**

22. $\frac{3 \text{ tsp.}}{\text{tbsp.}} \times \frac{2 \text{ tbsp.}}{\text{fl. oz.}} \times 3 \text{ fl. oz.} = $ **18 tsp.**

Part E

23. 78°F

25°C *or* 26°C

24. Answers will vary, but most people would choose *warm.*

Tables and Charts pp. 134–135

Part A

1. 3,005,000 people

2. 4

3. Los Angeles, CA

4. Philadelphia, PA

Part B

5. Mikki Yamomoto

6. Iris Russel

7. $560.50

8. William Hastings

Part C

9. 2,566 mi.

10. $2,537 - 2,312 = $ **225 mi.**

11. $1,883 + 2,566 = $ **4,449 mi.**

12. The empty boxes represent the distance from each city to itself.

Computer Spreadsheets pp. 136–137

Part A

1. $40 \times \$8.45 = $ **$338**

2. =B3*C3

3. $36 \times \$9.50 = $ **$342**

4. =D2+D3

Part B

5. =B2*.2
$37.50 \times .2 = $ **$7.50**

6. =B3*.25
$28.80 \times .25 = $ **$7.20**

7. =B2-C2
$37.50 - \$7.50 = $ **$30.00**

8. =B3-C3
$28.80 - \$7.20 = $ **$21.60**

9.

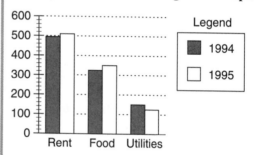

4	Shoes	$37.90	B4*.30 ($11.37)	B4-C4 ($26.53)
5	Shirt	$25.00	B5*.15 ($3.75)	B5-C5 ($21.25)
6	Sunglasses	$18.50	B6*.5 ($9.25)	B6-C6 ($9.25)

Bar Graphs pp. 138–139

Part A

1. Houston

2. Los Angeles

3. Estimates will vary. Should be between 2,500,000 and 3,000,000.

4. Answers will vary. You may have said to the nearest hundred thousand or the nearest million.

Part B

5. Houston, Los Angeles, New York

6. Chicago, Philadelphia

7. Los Angeles

8. Houston

Making Connections: Drawing a Bar Graph p. 139

Line Graphs pp. 140–141

Part A

1. about $2,000 billion (or $2 trillion)

2. 1980

3. 1970 to 1975

4. 1985 to 1990

Part B

5. U.S.

6. around 1964 or 1965

7. around 1977 or 1978

8. between 1960 and 1970

Part C

9. 30

10. between 1960 and 1970

11. from 1990 to 2000

12. Answers will vary. You may have mentioned people having fewer children, people having children at an older age, increasing life expectancy, post–World War II baby boomers passing peak childbearing age, and so on.

Circle Graphs pp. 142–143

Part A

1. 26.4%

2. 20.5 + 4 = **24.5%**

3. 100 − 20.5 = **79.5%**

4. increase

Part B

5. building construction

6. 11%

7. 13 + 10 = **23%**

8. 100 − 42 = **58%**

Part C

9. manufacturing (54%)

10. 9%

11. 80%

12. 265 × 11% ≈ **29**

Making Connections: Creating a Circle Graph p. 143

Williams Family Budget

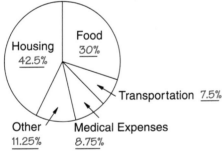

Scatter Diagrams pp. 144–145

Part A

Estimates will vary.

1. about $50,000

2. about $20,000

3. about $30,000 − $20,000 = **$10,000**

4. The more education a person has, the more likely that person is to earn a higher salary.

Part B

Commute Times

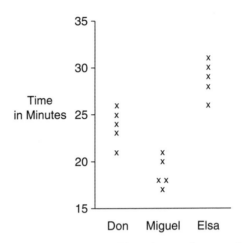

Making Connections: Choosing an Appropriate Graph p. 145

1. Circle graph. Each candidate's vote percent will be seen as a part of the entire vote.

   Bar graph. The percent of the vote for each candidate could be compared to opponents' percents.

2. Scatter diagram. Each child's data is represented. Clusters of dots will show relationship between grades and TV viewing.

3. Double line graph. Each city's temperatures are drawn independently, so the two can be compared.

4. Bar graph. The height of each bar shows the number of acres planted with each type of grain. Could also use a circle graph to see the percent of each grain relative to the total acreage.

Using More than One Data Source pp. 146–147

Part A

Your answers may vary but should be close to those shown here.

1. 963 mi. ÷ 22 mpg = 43.8 gal.

   43.8 gal. × $1.15 = **$50.37**

2. 479 mi. ÷ 38 mpg = 12.6 gal., rounded to **13 gal.**

3. 496 mi. ÷ 28 mpg × $\frac{\$1.20}{\text{gal.}}$ = $21.26 for compact car

   496 mi. ÷ 38 mpg × $\frac{\$1.20}{\text{gal.}}$ = $15.66 for subcompact car

   $21.26 − $15.66 = **$5.60**

4. $2{,}142$ mi. $\div 28$ mpg $\times \frac{\$1.10}{\text{gal.}} =$
$\$84.15$ for compact car
$2{,}142$ mi. $\div 22$ mpg $\times \frac{\$1.25}{\text{gal.}} =$
$\$121.70$ for midsize car
$\$121.70 - \$84.15 = \mathbf{\$37.55}$

Part B

5. $1{,}561{,}541 \times 36.2\% = \mathbf{565{,}278}$ (rounded)

6. $104{,}552{,}736 \times 43\% = \mathbf{44{,}957{,}676}$ (rounded)

7. larger (23.4% compared to 19%)

8. Clinton did worse in Colorado than in the United States as a whole.
Bush did almost the same in Colorado as in the United States as a whole.

Part C

9. Clinton got a higher percent in Adams County.

10. Jefferson and Arapahoe

Simple and Compound Probability pp. 148–149

Part A

1. 1 out of 7

2. 1 out of 2

3. 5 out of 6

Part B

4. 1 out of 4

5. 1 out of 13

6. 8 out of 52 $=$ 2 out of 13

7. 3 out of 52

Part C

8. 1 out of 4

9. 1 out of 8

10. 1 out of 36

11. There are 6 combinations that equal seven and 36 possible outcomes; 6 out of 36 $=$ 1 out of 6; $\frac{6}{36} = \frac{1}{6}$.

Seeing Trends, Making Predictions pp. 150–151

Part A

1. 4 white shirts out of 8 shirts total $=$
1 out of 2 *or* **50%**

2. There is a 54% chance that Tony is undecided.

3. 5 times

4. 14 bids

Part B

5. It takes fewer yen to buy a dollar. The relative value of the yen has been rising.

6. It takes more drachmas to buy a dollar. The relative value of the drachma has been falling.

7. The value of the drachma is dropping, so this is the most likely prediction.

Making Connections: Checking Probability p. 151

2 or 3 times

The more you roll the dice, the more accurate the prediction.

Mean, Median, and Mode pp. 152–153

Part A

1. 35 $\$1.77$ (rounded)   7 lb.

2. 168.6   7.4 oz. $\$48$ (rounded)

Part B

3. $\$23$ 22 $(13.4 + 8.6) \div 2 = \mathbf{11}$

Part C

4. 18 (3 occurrences)
22° (3 occurrences)

Part D

5. Answers will vary. Any reasonable answer is acceptable. Sample answers are given.

- An extreme value can greatly affect the value of a mean. Since there are no extreme values, a mean value would be a reasonable choice.

- In this problem the median value is 97, which is a reasonable typical value.

- The mode may be the best choice in representing the typical value since most of the scores cluster around 98.

Unit 4 Review pp. 154–155

Part A

1. Norris

2. 585 lb.

3. 33 to 34 years

4. Stern's age does not greatly affect the average age of the team. The median age is 32 compared to a mean of 33 to 34.

Part B

Estimates will vary, but your answers should be close to the ones shown here.

5. $635,000

6. second

7. ($525,000 + $375,000) ÷ 2 = **$450,000**

8. third

9. $525,000 + $375,000 + $900,000 + $625,000 = **$2,425,000**

10. $900,000 − $375,000 = **$525,000**

Part C

11. mortgage

12. =B2+B3

13. $323 (rounded to the nearest dollar)

14. $3,500 ($266 more than their expenses)

Part D

15. 31 out of 365
    31 out of 366

16. 1 out of 8

17. 15

18. 1 out of 2

19. 10 out of 20 = 1 out of 2

20. about $689

Working Together

Results will vary.

Unit 5

Writing Expressions pp. 158–159

Part A

1. $26 + 8$

2. $320 − $45

3. 15.6 + 2.8

4. 12 + 8

5. $195 − $50

Part B

6. $12 × 9 or $12(9) or $12 · 9

7. $3\overline{)\$450}$ or $\frac{\$450}{3}$ or $450/3 or $450 ÷ 3

8. 21 × 30 or 21(30) or 21 · 30

9. $5\overline{)75}$ or $\frac{75}{5}$ or 75/5 or 75 ÷ 5

10. $6.95 × 15 or $6.95(15) or $6.95 · 15

Part C

11. A + C

12. C − D

13. AB (You don't need parentheses to show the multiplication of two variables.)

14. C/D or $\frac{C}{D}$ or $D\overline{)C}$ or C ÷ D

The Number Line pp. 160–161

Part A

1. 10   17   −13   0

2. 22   6   0   2

Part B

3. −34   −27   5   −6

Part C

4. 17   −4   −3   35   7

Making Connections: Keeping Score p. 161

1. **160**

   90 − (−70) = 160

2. **3 hands**

   500 − 220 = 280

   $\frac{280}{120}$ = 2 R40

3. Robin −120

   Art 150

   Colin 190

   Cassady 50

   −70 + (−50) = −120

   90 + 60 = 150

   220 − 30 = 190

   −30 + 80 = 50

Powers and Roots pp. 162–163

Part A

1. 49   27   8

2. 121   1   144

3. 125   64   729

4. 16   216   1,000

Part B

$2^2 = 4$	$5^2 = 25$	$8^2 = 64$	$11^2 = 121$	$14^2 = 196$
$3^2 = 9$	$6^2 = 36$	$9^2 = 81$	$12^2 = 144$	$15^2 = 225$
$4^2 = 16$	$7^2 = 49$	$10^2 = 100$	$13^2 = 169$	$20^2 = 400$

Part C

5. 10   11   7   8

6. 12   13   20   9

Part D

7. **(3) 7 and 8**

$7 \cdot 7 = 49$

$8 \cdot 8 = 64$

50 is between 49 and 64.

8. **(2) 11 and 12**

$11 \cdot 11 = 121$

$12 \cdot 12 = 144$

125 is between 121 and 144.

9. **(1) 3 and 4**

$3 \cdot 3 = 9$

$4 \cdot 4 = 16$

10 is between 9 and 16.

10. Your answer should include some of the ideas from this sample answer:

"To make an estimate, I thought of a perfect square that I knew that was near 90. $10 \cdot 10 = 100$. Because 100 is greater than 90, I tried a lower number. $9 \cdot 9 = 81$. Since 81 is less than 90 and 100 is greater than 90, I know that the square root of 90 must be between 9 and 10."

Writing Equations pp. 164–165

Part A

1. **(2)** $x - 12 = 25$

2. **(1)** $\frac{60}{n} = 15$

3. **(3)** $a + 9 = 20$

4. **(1)** $3w - 2 = 10$

5. **(2)** $5b + 2 = 12$

6. **(1)** $\frac{10}{z} = 3 - z$

Part B

Answers may vary, but should be similar to these sample answers.

7. The sum of five and a number is eight.

8. Twelve divided by a number is six.

9. Five less than four times a number is seven.

Part C

10. **(2)** $\$320 + d = \$400$

11. **(3)** $4s = 12$

12. **(2)** $p - \$35 = \$85$

13. **(1)** $\$1.15n = \$5.75$

14. **(1)** $5d = 1{,}650$

15. Both equations mean the same thing. It doesn't matter which side of the equation the expressions are on as long as the sides are equal.

Order of Operations pp. 166–167

Part A

1. 26   5   44

2. 3   12   405

3. 4   45   1

4. 1   4   5

5. 121   12   81

Part B

6. **(2)** $42 - 20$

7. **(1)** $\frac{60}{10} - 5$

8. **(3)** $2(3)$

9. **(2)** $\frac{39}{3}$

10. **(1)** $13(3)$

11. **(3)** $15 + 6$

Part C

12. $\dfrac{7(\$9) + 6(\$12)}{3}$

You may have set the problem up differently, but your answer should include these elements: multiplying $9 by 7, multiplying $12 by 6, adding these products, and dividing the sum by 3.

13. The order of operations is important in solving this problem. If you add before multiplying, you will get an incorrect answer.

The Distributive Property pp. 168–169

Part A

1. **(2)** $3 \cdot 8 - 3 \cdot 5$

2. **(1)** $c(15 - 10)$

3. **(1)** $9(5 + 6)$

4. **(3)** $6x + 6 \cdot 12$

Part B

5. $2(\$550) + 2(\$28)$ or $2(\$550 + \$28)$

6. $40(\$16) - 40(\$11)$ or $40(\$16 - \$11)$

7. $3(6) - 3(3)$ or $3(6 - 3)$

8. $4(0.7) + 4(4.5)$ or $4(0.7 + 4.5)$

Making Connections: Perimeter of a Rectangle p. 169

1. $P = 2l + 2w$

   $P = 2(11) + 2(8.5)$

   $P = 22 + 17$

   $P =$ **39 centimeters**

2. $P = 2(l + w)$

Addition and Subtraction Equations pp. 170–171

Part A

1. $a - 15 = 78$

   $a - 15 + 15 = 78 + 15$

   $a =$ **93**

   $5.3 = x + 0.6$

   $5.3 - 0.6 = x + 0.6 - 0.6$

   4.7 $= x$

   $532 = 189 + n$

   $532 - 189 = 189 + n - 189$

   343 $= n$

2. $b + 3\frac{1}{2} = 9$

   $b + 3\frac{1}{2} - 3\frac{1}{2} = 9 - 3\frac{1}{2}$

   $b =$ **$5\frac{1}{2}$**

   $10 = w - 8.2$

   $10 + 8.2 = w - 8.2 + 8.2$

   18.2 $= w$

   $d - 11 = 33$

   $d - 11 + 11 = 33 + 11$

   $d =$ **44**

3. $x + 6.4 = 100$

   $x + 6.4 - 6.4 = 100 - 6.4$

   $x =$ **93.6**

   $1,050 + g = 8,000$

   $1,050 + g - 1,050 = 8,000 - 1,050$

   $g =$ **6,950**

   $9\frac{1}{2} = y - 3\frac{1}{2}$

   $9\frac{1}{2} + 3\frac{1}{2} = y - 3\frac{1}{2} + 3\frac{1}{2}$

   13 $= y$

Part B

4. $c + 28 - 7 = 84$

   $c + 21 = 84$

   $c + 21 - 21 = 84 - 21$

   $c =$ **63**

   $h + 1.5 + 8.75 = 12.5$

   $h + 10.25 = 12.5$

   $h + 10.25 - 10.25 = 12.5 - 10.25$

   $h =$ **2.25**

5. $95 - 25 + 15 + x = 105$

   $85 + x = 105$

   $85 + x - 85 = 105 - 85$

   $x =$ **20**

   $5.375 = e - 0.9 + 2.05$

   $5.375 = e + 1.15$

   $5.375 - 1.15 = e + 1.15 - 1.15$

   4.225 $= e$

Part C

6. $265 + m = 770$

   $m =$ **505 miles**

7. $\$420 + c = \$795$

   $c =$ **\$375**

8. $2.8 + 2.1 + a = 6.5$

   $4.9 + a = 6.5$

   $a =$ **1.6 centimeters**

9. You need to subtract 24 degrees from the high temperature. Since the high temperature, 18, is less than 24, you know the resulting low temperature will be a negative number.

   $18 - 24 = l$

   $-6 = l$

   The low temperature was **−6°F.**

Multiplication and Division Equations pp. 172–173

Part A

1. $\frac{w}{6} = 9$

   $6 \cdot \frac{w}{6} = 6 \cdot 9$

   $w =$ **54**

   $12n = 600$

   $\frac{12n}{12} = \frac{600}{12}$

   $n =$ **50**

   $\frac{z}{16} = 128$

   $16 \cdot \frac{z}{16} = 16 \cdot 128$

   $z =$ **2,048**

2. $1.5x = 45$

   $\frac{1.5x}{1.5} = \frac{45}{1.5}$

   $x =$ **30**

   $\frac{p}{2} = 56$

   $2 \cdot \frac{p}{2} = 2 \cdot 56$

   $p =$ **112**

   $25r = 40$

   $\frac{25r}{25} = \frac{40}{25}$

   $r =$ **1.6**

3. $50c = 3,000$

$\frac{50c}{50} = \frac{3,000}{50}$

$c = \mathbf{60}$

$\frac{y}{0.5} = 150$

$0.5 \cdot \frac{y}{0.5} = 0.5 \cdot 150$

$y = \mathbf{75}$

$\frac{h}{32} = 3$

$32 \cdot \frac{h}{32} = 32 \cdot 3$

$h = \mathbf{96}$

Part B

4. $7x + 3 = 17$

$7x = 14$

$x = \mathbf{2}$

$\frac{y}{6} - 3 = 1$

$\frac{y}{6} = 4$

$y = \mathbf{24}$

$10x - 6 = 24$

$10x = 30$

$x = \mathbf{3}$

5. $3b + 7 = b - 9$

$2b + 7 = -9$

$2b = -16$

$b = \mathbf{-8}$

$4n - 15 = 7n - 51$

$4n + 36 = 7n$

$36 = 3n$

$\mathbf{12 = n}$

$\frac{z}{3} - 4 = 2$

$\frac{z}{3} = 6$

$z = \mathbf{18}$

Part C

6. $\$200 + \$15a = \$470$

$\$15a = \$270$

$a = 18$

Stuart sold **18 ads** last week.

7. $2h + 4 = 38$

$2h = 34$

$h = 17$

John worked **17 hours** last week.

8. $3x + x = 60$

$4x = 60$

$x = 15$

They lost **15 games.**

9. $36 + 36 + w + w = 96$

$72 + 2w = 96$

$2w = 24$

$w = 12$

The rectangle is **12 inches wide.**

10. You can use the formula as the equation. Put in the information from the problem and solve.

$P = 2l + 2w$

$96 = 2(36) + 2w$

$96 = 72 + 2w$

$24 = 2w$

$12 = w$

Working with Formulas pp. 174–175

Part A

1. a. $A = lw$

$\frac{A}{l} = w$

b. $\frac{180}{15} = w$

$\frac{180}{15} = \mathbf{12\ inches}$

2. a. $P = a + b + c$

$P - a - b = c$

b. $72 - 20 - 18 = c$

34 centimeters $= c$

3. a. $a = 2x - b$

b. $2(65) - 50 = a$

$130 - 50 = a$

$\mathbf{80} = a$

Making Connections: The Distance Formula p. 175

1. $d = rt$

$d = 20\left(2\frac{1}{2}\right)$

$d = \mathbf{50\ miles}$

2. $r = \frac{d}{t}$

$r = \frac{60}{12}$

$r = \mathbf{5\ feet\ per\ second}$

Substituting to Solve Equations pp. 176–177

Part A

1. $3x - y = 12$

$3x - 6 = 12$

$3x = 18$

$x = \mathbf{6}$

2. $2m + 2n = 10$

$2(3) + 2n = 10$

$6 + 2n = 10$

$2n = 4$

$n = \mathbf{2}$

3. $5a + 2c = -12$
   $5a + 2(-1) = -12$
   $5a - 2 = -12$
   $5a = -10$
   $a = -2$

4. $4(s + 4t) = 100$
   $4(5 + 4t) = 100$
   $20 + 16t = 100$
   $16t = 80$
   $t = 5$

Part B

5. $3r = r - 8$
   $2r = -8$
   $r = -4$
   $s = -2r$
   $s = -2(-4)$
   $s = 8$

6. $7w + 10 = 80$
   $7w = 70$
   $w = 10$
   $2v = 3w$
   $2v = 3(10)$
   $2v = 30$
   $v = 15$

7. $10c - 6 = 2(c + 1)$
   $10c - 6 = 2c + 2$
   $8c = 8$
   $c = 1$
   $2d = 8c$
   $2d = 8(1)$
   $2d = 8$
   $d = 4$

8. $9(x + 4) = 4x - 39$
   $9x + 36 = 4x - 39$
   $5x = -75$
   $x = -15$
   $45 - 2x = y$
   $45 - 2(-15) = y$
   $45 + 30 = y$
   $75 = y$

Part C

9. $2a + 7b = 78$
   $2(b - 6) + 7b = 78$
   $2b - 12 + 7b = 78$
   $9b = 90$
   $b = 10$

   $a = b - 6$
   $a = 10 - 6$
   $a = 4$

10. $4e + 2d = -6$
    $4(d + 9) + 2d = -6$
    $4d + 36 + 2d = -6$
    $6d = -42$
    $d = -7$

    $e = d + 9$
    $e = -7 + 9$
    $e = 2$

Part D

11. **Tom earns \$5.25 per hour.**

    Solve for m.
    $40m + \$16 = \$436$
    $40m = \$420$
    $m = \$10.50$

    Substitute.
    $m = 2t$
    $\$10.50 = 2t$
    $\$5.25 = t$

12. **80 miles**

    Substitute.
    $v - b = 160$
    $3b - b = 160$
    $2b = 160$
    $b = 80$

13. **length = 100; width = 30**

    Write an equation, substitute 20 for y, and solve for x.
    You can use the formula for perimeter of a rectangle.
    $P = 2l + 2w$
    $260 = 2(5y) + 2(x + y)$
    $260 = 2(5)(20) + 2(x + 20)$
    $260 = 200 + 2x + 40$
    $260 = 240 + 2x$
    $20 = 2x$
    $10 = x$

    Now return to the original expressions.
    length $= 5y = 5(20) = 100$
    width $= x + y = 10 + 20 = 30$

250

14. Yes. The length and width of the rectangle are expressed using variables. You need to know the perimeter in order to set up an equation. Without the perimeter, you know some things about the relationship between the variables, but you cannot solve for their values.

Writing and Solving Inequalities pp. 178–179

Part A

1. $4b - 3 \geq 9$

$4b \geq 12$

$b \geq 3$

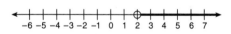

$2(x + 5) \leq 18$

$2x + 10 \leq 18$

$2x \leq 8$

$\boldsymbol{x \leq 4}$

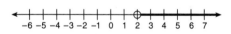

$12 > 6m + 2 - m$

$12 > 5m + 2$

$10 > 5m$

$\boldsymbol{2 > m}$

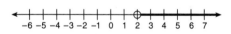

2. $2n + 8 > 4$

$2n > -4$

$\boldsymbol{n > -2}$

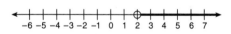

$6 \leq 3(y - 2)$

$6 \leq 3y - 6$

$12 \leq 3y$

$\boldsymbol{4 \leq y}$

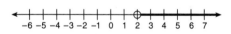

$7 < 3a - 1 + a$

$7 < 4a - 1$

$8 < 4a$

$\boldsymbol{2 < a}$

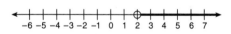

Part B

3. **(3)** 7 **(4)** 5 **(5)** 2 **(6)** −1

$y - 4 \leq 3$

$y \leq 7$

4. **(1)** 3 **(2)** 1 **(3)** 0

$b + 2 > 1$

$b > -1$

5. **(4)** 4 **(5)** 8 **(6)** 12

$2x - 10 \geq -2$

$2x \geq 8$

$x \geq 4$

Part C

6. $x + 10 \geq 6$

$\boldsymbol{x \geq -4}$

7. $x + 5 > -3$

$\boldsymbol{x > -8}$

8. $x - 4 < 2x - 10$

$x + 6 < 2x$

$\boldsymbol{6 < x}$

9. $2x + 18 \leq 8x$

$18 \leq 6x$

$\boldsymbol{3 \leq x}$

Part D

10. **6 Super Burgers with cheese**

Think: Some number of Super Burgers times $2.19 is less than or equal to $15.

$\$2.19x \leq \$15$

$x \leq \dfrac{\$15}{\$2.19}$

$x \leq 6 \text{ R}\$1.86$

11. **4 meals**

$\$1.49x + \$0.59x + \$1.09x \leq \$15$

$\$3.17x \leq \$15$

$x \leq \dfrac{\$15}{\$3.17}$

$x \leq 4 \text{ R}\$2.32$

12. **3 Junior Meals**

$\$2.19 + \$0.99 + \$1.09 = \$4.27$ (the cost of Frank's meal)

$\$15 - \$4.27 = \$10.73$ (the amount left over)

$\$2.99x \leq \$10.73$

$x \leq \dfrac{\$10.73}{\$2.99}$

$x \leq 3 \text{ R}\$1.76$

13. **Cost of 1 gallon of soda $<$ Cost of 8 large drinks**

$8(\$1.09) = \$8.72$

$\$5 < \$8.72$

Translating Words to Equations pp. 180–181

Part A

1. $18

Parking	Speeding
x	3x + $6

$x + 3x + \$6 = \$78$
$4x + \$6 = \$78$
$4x = \$72$
$x = \$18$

2. 36 hours

Al	Art	Anna
x	2x − 6	2x − 6 + 10

$x + 2x - 6 + 2x - 6 + 10 = 103$
$5x - 12 + 10 = 103$
$5x - 2 = 103$
$5x = 105$
$x = 21$
$2x - 6 = 2(21) - 6 = 42 - 6 = 36$

Part B

3. 8 years old

$5x + 10 + x + 10 = 68$
$6x + 20 = 68$
$6x = 48$
$x = 8$

4. 2 years old

	Chris	Ann
Age Now	x + 4	x
In 2 Years	x + 4 + 2	x + 2

Think: Chris's age in 2 years will be equal to twice Ann's age in 2 years.

$x + 4 + 2 = 2(x + 2)$
$x + 6 = 2x + 4$
$2 = x$

Part C

5. 8 dimes and 16 nickels

	Dimes	Nickels
Number of Coins	x	2x
Value	10x	5(2x)

Since you are using cents to show the value of the coins, express $1.60 as 160 cents.

$10x + 5(2x) = 160$
$10x + 10x = 160$
$20x = 160$
$x = 8$ (dimes)
$2x = 16$ (nickels)

6. 5 of each bill

	$1 bills	$5 bills	$10 bills
Value	1x or x	5x	10x

$x + 5x + 10x = 80$
$16x = 80$
$x = 5$

7. 20 quarters

	Dimes	Quarters
Number of Coins	x	4x
Value	10x	25(4x)

$10x + 25(4x) = 550$
$10x + 100x = 550$
$110x = 550$
$x = 5$ dimes

Now substitute the value for x in $4x$ (the number of quarters).

$4(5) = 20$ quarters

8. Answers will vary. Here is a sample problem.

A cash register drawer contains $525 in $20 bills, $10 bills, and $5 bills. If there are the same number of each bill, how many are there of each?

Sample chart:

$20s	$10s	$5s

Mixed Review pp. 182–183

Part A

1. $6 \cdot \$8$ *or* $6(\$8)$

2. $\$78/3$ *or* $3\overline{)\$78}$ *or* $\frac{\$78}{3}$ *or* $\$78 \div 3$

3. $\$5 + \$50$

4. $\$550 - \$100$

Part B

5. 16

6. 27

7. 169

Part C

8. **(1)** 20

9. **(3) between 16 and 17**

   $16^2 = 256$

   $17^2 = 289$

10. **(1) between 5 and 6**

    $5^2 = 25$

    $6^2 = 36$

Part D

11. $-5 < -1$

12. $+6 = 6$

13. $2 > -2$

Part E

14. 6

15. -16

16. 18

17. -25

18. 0

19. 125

Part F

20. $n - \$5.85$

21. $2h - 10$

22. $\frac{x}{y}$

23. $3v + 3{,}500$

Part G

24. $4x - 5 = 7$

    $4x = 12$

    $\textbf{\textit{x}} = \textbf{3}$

25. $12 - 4y = -4$

    $-4y = -16$

    $\textbf{\textit{y}} = \textbf{4}$

26. $3x - 1 > 6x - 7$

    $6 > 3x$

    $\textbf{2} > \textbf{\textit{x}}$

27. $4a - 9 = 8a + 3$

    $-12 = 4a$

    $\textbf{--3} = \textbf{\textit{a}}$

28. $6n + 15 \geq 8n - 1$

    $16 \geq 2n$

    $\textbf{8} \geq \textbf{\textit{n}}$

29. $3z - 8 < 10z - 2 - 8z$

    $3z - 8 < 2z - 2$

    $\textbf{\textit{z}} < \textbf{6}$

Part H

30. $5x^2 - y = 20$

    $5(25) - y = 20$

    $125 - y = 20$

    $\textbf{105} = \textbf{\textit{y}}$

31. $4a - 12 = 0$

    $4a = 12$

    $\textbf{\textit{a}} = \textbf{3}$

    $b = 2a + 4$

    $b = 2(3) + 4$

    $\textbf{\textit{b}} = \textbf{10}$

32. $-4c + 2 = 2(c - 8)$

    $-4c + 2 = 2c - 16$

    $18 = 6c$

    $\textbf{3} = \textbf{\textit{c}}$

    $\frac{1}{3}d = 3c$

    $\frac{1}{3}d = 3(3)$

    $\frac{1}{3}d = 9$

    $\textbf{\textit{d}} = \textbf{27}$

33. $5m + 3n = 26$

    $5(n + 2) + 3n = 26$

    $5n + 10 + 3n = 26$

    $8n = 16$

    $\textbf{\textit{n}} = \textbf{2}$

    $m = n + 2$

    $m = 2 + 2$

    $\textbf{\textit{m}} = \textbf{4}$

Part I

Problems 34–36 can be solved by writing equations with one variable or by writing two-variable equations and then using substitution to solve. Either method is correct. The substitution method is shown here.

34. 37 and 47 inches

$x + y = 84$ and $y = x + 10$

$x + x + 10 = 84$

$2x + 10 = 84$

$2x = 74$

$x = 37$

$y = 37 + 10$

$y = 47$

35. \$40

Let $x = $ November's bill

Let $y = $ July's bill

$x = 2y + 8$ and $x + y = 56$

Substitute:

$x + y = 56$

$2y + 8 + y = 56$

$3y = 48$

$y = 16$

$x = 2y + 8$

$x = 2(16) + 8$

$x = 32 + 8$

$x = 40$

36. Class A = 36 students; Class B = 24 students; Class C = 25 students

Let $x = $ the number in Class B.

Let $x + 12 = $ the number in Class A.

Let $y = $ the number in Class C.

We can write these equations:

$x + x + 12 + y = 85$ and $y = 2x - 23$

$x + x + 12 + 2x - 23 = 85$

$4x - 11 = 85$

$4x = 96$

$x = 24$ (Class B)

Class A $= x + 12 = 24 + 12 = 36$

$y = 2(24) - 23$

$y = 48 - 23$

$y = 25$ (Class C)

Points, Lines, and Angles pp. 184–185

Part A

1. (2)   3. (2)

2. (1)   4. (1)

Part B

5. **d**, an acute angle

   b, a right angle

   c, an obtuse angle

   a, a straight angle

6. (1) because the opening between the rays is smallest

7. Answers will vary. Here are some sample answers:

   perpendicular lines: dividers between windowpanes; lines formed by square tiles

   parallel lines: venetian blinds; bookshelves

   right angles: corners of square tiles, windowpanes, doors; angle of wall to floor

   acute angle: TV antenna

   obtuse angle: angle of door to wall when door is partially opened

   straight angle: angle of door to wall when door is closed

Protractors pp. 186–187

Part A

1. a. 75°   2. b. 180°

   b. 60°   a. 90°

   c. 110°   c. 30°

   d. 150°   d. 135°

Part B

3. **72°**

   $160° - 88° = 72°$

4.

Angle	Type	Degrees
∠POS	obtuse	125°
∠QOS	acute	80°
∠POR	acute	70°

5. ∠d = 30° and ∠c = 90°

   Let $x = $ ∠d

   $3x + x = 120°$

   $4x = 120°$

   $x = \mathbf{30°}$

   $3x = \mathbf{90°}$

6. **24**

   The hands form a straight angle once per hour.

Types of Angles pp. 188–189

Part A

1. 56° 5° 28° 81° 45°

Part B

2. 135° 90° 65° 15° 58°

Part C

3. ∠x and ∠z; ∠w and ∠y

4. ∠x and ∠z

5. **70°**

   180° − 110° = 70°

6. **110°**

   ∠x and ∠z are vertical angles.

Part D

7. ∠MOQ

8. **30°**

   ∠MOP + ∠POQ = 90°

   90° − ∠MOP = ∠POQ

   90° − 60° = 30°

9. 6 angles

   ∠NOM, ∠MOP ∠NOP, ∠NOQ, ∠MOQ, and ∠POQ

Making Connections: City Planning p. 189

1. Answers will vary. Make sure you have a logical reason for the road you are proposing. All possibilities have advantages and disadvantages.

2. Regardless of your proposed road, you have created supplementary angles, since the two parallel streets can be considered straight (180°) angles. If you proposed a perpendicular roadway, you created two equal 90° angles, which are supplementary.

Circles pp. 190–191

Part A

1. diameter

2. radius

3. They are equal in length. Because O marks the midpoint of the circle, OX is also a radius.

4. **14 inches**

   2(7) = 14

5. **15 centimeters**

   $\frac{30}{2}$ = 15

Part B

6. **10.4 inches**

   $C = \pi d$

   $C = 3.14(3)$

   $C = 9.42$

   $9.42 + 1 = 10.42$

   10.42 rounds to 10.4

7. **8.1 inches**

   $d = 2r$

   $d = 2(1.25)$

   $d = 2.5$

   $C = \pi d$

   $C = 3.14(2.5)$

   $C = 7.85$

   $7.85 + .25 = 8.1$

8. **9.6 feet**

   $C = \pi d$

   $30 = 3.14d$

   $\frac{30}{3.14} = d$

   $9.6 \approx d$

9. **59.2 centimeters**

   $C = \pi d$

   $C = 3.14(9)$

   $C = 28.26$

   $28.26 \cdot \frac{3}{4} \approx 21.2$

   Hint: If you are using a calculator, use 0.75 for $\frac{3}{4}$.

   The circumference of the circular portion is about 21.2 centimeters

   $P = 21.2 + 9 + 10 + 10 + 9 = 59.2$

Making Connections: Partitioning p. 191

Discussion will vary.

1. It is possible to cut a circle into 7 pieces with 3 straight lines.

2. You need at least 4 straight lines to cut the circle into 11 pieces.

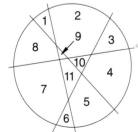

255

Quadrilaterals pp. 192–193

Part A

1. BD
2. AB and CD
3. CB and AD
4. ∠CDB

Part B

5. The opposite sides are equal and parallel; all angles are right angles; and the diagonals are equal.

6. **True.** The main property of a parallelogram, which is that the opposite sides are equal and parallel, is true of a square, a rectangle, and a rhombus.

7. A rhombus

8. **False.** A trapezoid has only one pair of parallel sides. Therefore, opposite angles cannot be equal. A trapezoid can have one right angle, but that angle's opposite cannot be a right angle.

Part C

9. No. Since both pairs of opposite sides are equal, the opposite sides must also be parallel. Given these conditions, the opposite angles must be equal. If there is one right angle, all angles must measure 90°.

10. A square. Since the adjacent sides XW and WZ are equal, the figure must be a square.

11. Yes for both. See the examples below.

Triangles pp. 194–195

Part A

1. right triangle
2. equilateral triangle
3. isosceles triangle
4. scalene triangle

Part B

5. Choices **(2)**, **(3)**, and **(5)** could form triangles. Total the angles. Only those that total 180° can form a triangle.

6. Side CD, because it is opposite the largest angle.

7. Angle MNO, because it is opposite the longest side.

Part C

8. 40°

9. ∠A is **20°** and ∠C is **40°**

$$120 + x + 2x = 180$$
$$3x = 60$$
$$x = 20, 2x = 40$$

10. Each measures **65°**

You know that ∠V and ∠W are equal, so let x represent the measure of each.

$$x + x + 50 = 180$$
$$2x + 50 = 180$$
$$2x = 130$$
$$x = 65°$$

11. $180° + 180° = 360°$

Part D

12. ∠D

13. Yes. The other angles must each measure 45°. The lengths of the legs must be equal although they can be any length.

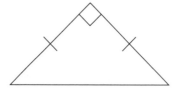

Similar Geometric Figures pp. 196–197

Part A

1. **Yes.** Both figures have the same shape, and their cross products are equal.

2. **Yes.** Both figures have the same shape because both are isosceles right triangles.

3. **No.** The ratio of length to width is not the same for both rectangles. $\frac{10}{4}$ does not equal $\frac{5}{3}$.

4. **Yes.** Although the triangles are in different positions, the corresponding sides have the same ratio. The lengths of the sides on the shorter triangle are doubled in the larger triangle.

Part B

5. $x = 40$ **and** $y = 24$

$\frac{x}{24} = \frac{30}{18}$

$30 \cdot 24 = 18x$

$720 = 18x$

$40 = x$

$\frac{y}{18} = \frac{32}{24}$

$32 \cdot 18 = 24y$

$576 = 24y$

$24 = y$

6. $4\frac{1}{2}$ **inches**

Set up a proportion:

$\frac{3}{5} = \frac{x}{7\frac{1}{2}}$

Hint: If you are using a calculator, use 7.5 for $7\frac{1}{2}$.

$3 \cdot 7.5 = 5x$

$22.5 = 5x$

$4.5 = x$

7. **26 feet**

$\frac{4}{6} = \frac{x}{39}$

$4 \cdot 39 = 6x$

$156 = 6x$

$26 = x$

8. **140 yards**

$\frac{1.75 \text{ cm}}{2.5 \text{ cm}} = \frac{w}{200 \text{ yd.}}$

$1.75 \cdot 200 = 2.5w$

$350 = 2.5w$

$140 = w$

9. $4\frac{1}{2}$ **or 4.5**

Set up a proportion:

$\frac{XY}{YZ} = \frac{VX}{VW}$

$\frac{12}{9} = \frac{6}{x}$

$9 \cdot 6 = 12x$

$54 = 12x$

$4.5 \text{ or } 4\frac{1}{2} = x$

10. **78 feet**

$\frac{6}{5} = \frac{x}{65}$

$6(65) = 5x$

$390 = 5x$

$78 = x$

11. **a. True.** Since the sides in any equilateral triangle are equal, the ratio of the sides must always be the same.

  b. True. Since the sides are always equal in any square, the sides must always maintain the same ratio.

c. False. The lengths of the sides may not maintain the same ratio.

d. False. The lengths of the sides may not maintain the same ratio.

Finding Patterns in Algebra and Geometry
pp. 198–199

Part A

1. **18**

Each number is 3 more than the preceding number.

2. **55**

Each number is 9 less than the preceding number.

3. **10,000**

Each number is 10 times the preceding number.

4. **32**

Each number is 2 times the preceding number.

5. **95**

Each number is 2 times the preceding number, plus 1.

6. $\frac{1}{16}$

Each number is one half the preceding number.

Part B

7. **7:00**

Each clock is 1 hour 15 minutes later than the preceding one.

8. **U**

A = 1, E = 5, I = 9, M = 13, Q = 17, U = 21

The position of each letter in the alphabet is 4 more than the preceding letter.

Part C

Rectangle	Length	Width	Perimeter
ABCD	2	1	6
EFGH	3	2	10
IJKL	4	3	14
MNOP	5	4	18
QRST	6	5	22
UVWX	7	6	26

9. 22

10. 8

Part D

Number of steps	2	3	4	5	6
Number of cubes	3	6	10	15	21

11. 36

12. 20

The number of steps is the number of squares in the bottom row. Find the pattern by comparing the number of steps and the number of squares on the bottom row in the diagram.

The Pythagorean Theorem pp. 200–201

Part A

1. **26 inches**

$$10^2 + 24^2 = c^2$$
$$\sqrt{100 + 576} = c$$
$$\sqrt{676} = c$$
$$26 = c$$

2. **24 centimeters**

$$25^2 - 7^2 = b^2$$
$$\sqrt{625 - 49} = b$$
$$\sqrt{576} = b$$
$$24 = b$$

3. **41 centimeters**

$$9^2 + 40^2 = c^2$$
$$\sqrt{81 + 1,600} = c$$
$$\sqrt{1,681} = c$$
$$41 = c$$

4. **82 feet**

$$18^2 + 80^2 = c^2$$
$$\sqrt{324 + 6,400} = c$$
$$\sqrt{6,724} = c$$
$$82 = c$$

Part B

5. You need to know the length and width of one rectangle. The diagonal divides the rectangle into two equal right triangles. The diagonal is the hypotenuse.

Then you could use the formula $a^2 + b^2 = c^2$, where a = the length and b = the width and c = the hypotenuse (or diagonal).

The length of the diagonals, which will be covered by border tile separating the blue and yellow tiles, would be 4 times the hypotenuse.

6. The surface area of the two tiled portions will be the same because each diagonal divides its rectangle into two equal halves.

Reading Maps pp. 202–203

Part A

1. a. 2 inches = **16 miles**

   b. 1 inch = **8 miles**

   c. $3\frac{1}{2}$ inches = **28 miles**

   d. $2\frac{1}{2}$ inches = **20 miles**

   e. 5 inches = **40 miles**

2. **13 miles**

You can find the diagonal distance using the Pythagorean theorem or by measuring.

The two legs of the right triangle are 8 miles and 10 miles.

$$8^2 + 10^2 = c^2$$
$$\sqrt{64 + 100} = c$$
$$\sqrt{164} = c$$

c is between 12 and 13 miles. Since the square of 13 is 169, the distance is closer to 13 miles.

The measure of the diagonal is $1\frac{5}{8}$ inches on the map, which equals 13 miles.

3. **$27\frac{1}{2}$ miles; yes, the road would pass directly through Coalville.**

The distance on the map is $3\frac{7}{16}$ inches, which equals $27\frac{1}{2}$ miles.

$$3\tfrac{7}{16} \cdot 8 = 24\tfrac{56}{16} = 24 + 3\tfrac{1}{2} = 27\tfrac{1}{2}$$

Part B

4. **(3) 1,300 feet**

The distance of about $4\frac{1}{2}$ inches = 1,350 feet

$$4\tfrac{1}{2} \cdot 300 = 1,200 + 150 = 1,350$$

5. **830 feet**

The distance of $2\frac{3}{4}$ inches = 825 feet

$$2\tfrac{3}{4} \cdot 300 = 600 + 225 = 825, \text{ rounded to 830 feet}$$

6. **(2) 1,800 feet**

The distance is about 6 inches, which equals 1,800 feet.

7. Your directions may vary. A sample answer follows: "Make a left on Ogden; at the end of the street turn left on Hauser; go half a block and turn right on Fourth Street, then left on La Brea. Go past the shopping center and turn left on Third Street by the bank. The tennis courts will be on the left side of the street."

Perimeter and Circumference pp. 204–205

Part A

1. 60 meters

2. 251.2 feet

3. 27 inches

4. 25.12 centimeters

5. 11 inches

6. 648 miles

Part B

7. **54**

   $x = 12$ and $y = 15$

8. **40**

   $s = 10$ and $t = 4$

9. **40**

   $a = 7, b = 10,$ and $c = 7$

Part C

10. 52 feet

11. **83.1 feet**

    Find the circumference of the circle and add it to the two sides of the rectangle.

12. 28.3 feet

13. To find the measure of side a use the Pythagorean theorem, with 9 and 12 as the legs of the right triangle. To find the measure of side b add 9 and 3.

Area pp. 206–207

Part A

1. **5.5 square inches**

   Use the formula for the area of a parallelogram.

2. **28 square feet**

   Use the formula for the area of a rectangle.

3. **78.5 square centimeters**

   Use the formula for the area of a circle.

4. **169 square feet**

   Use the formula for the area of a square.

5. **300 square meters**

   Use the formula for the area of a triangle.

Part B

6. **180 square feet**

   $A = lw$

   $A = 15(12)$

   $A = 180$

7. **254.34 square feet**

   $A = \pi r^2$

   $A = 3.14(9)^2$

   $A = 3.14(81)$

   $A = 254.34$

Part C

8. **126 square centimeters**

   $12(9) + 3(6) = 108 + 18 = 126$

9. **320 square inches**

   $20^2 - \frac{1}{2}(20)(8) = 400 - 80 = 320$

10. **272.52 square feet**

    $12(18) + .5(3.14)(6)^2 = 216 + 56.52 = 272.52$

Part D

11. **880 square feet**

    $32^2 - \frac{1}{2}(18)(16) = 1024 - 144 = 880$

12. **Not necessarily.**

    Dimensions and perimeter of rectangles with an area of 24 square units:

Length	Width	Perimeter
6	4	20
8	3	22
12	2	28
24	1	50

Volume pp. 208–209

Part A

1. **96 cubic inches**

   Use the formula for volume of a rectangular solid.

2. **27 cubic centimeters**

   Use the formula for volume of a cube.

3. **197.8 cubic centimeters**

   Use the formula for volume of a cylinder.

4. **24,617.6 cubic inches**

   Use the formula for volume of a cylinder.

5. **336 cubic feet**

   Use the formula for volume of a rectangular solid.

6. **300 cubic feet**

   Use the formula for volume of a rectangular solid.

Part B

7. 72 cubic feet

8. 35 cubic feet

First box: 8(5)(4) = 160 ft.^{3}
Second box: 5^3 = 125 ft.^{3}
160 − 125 = 35

9. 4,019.2 cubic feet

If the diameter is 16 feet, the radius is 8 feet.
$V = 3.14(8)^2(20)$
$V = 3.14(64)(20)$
$V = 4,019.2$

10. a. 144 cubic feet

b. $5\frac{1}{3}$ cubic yards

11. No; they don't have equal volume. The shorter cylinder has a volume of about 29.4 cubic inches. The taller cylinder has a volume of about 25.1 cubic inches.

Although the pieces of paper have the same area, their cylinders do not have the same volume. The difference comes from the differences in surface area of the ends of the cylinders, which are not formed by the paper.

Choosing Area, Perimeter, or Volume pp. 210–211

Part A

1. a. perimeter

b. $P = 2l + 2w$
2(60) + 2(72) = 120 + 144 = **264 inches**

2. a. volume

b. $V = lwh$
$V = 15 \cdot 7 \cdot 7$
$V =$ **735 cubic feet**

3. a. area

b. $A = \pi r^2$
$A = 3.14(18)^2$
$A = 3.14(324)$
$A =$ **1,017.36 square feet**

4. a. area

b. $A = lw$
$A = 22(15) = 330$ square feet

Since each tile is 1 square foot, the school will need 330 rubber tiles.

Part B

5. a. feet

b. $2(4) + 2\left(2\frac{1}{2}\right) = 8 + 5 =$ **13 feet**

6. a. cubic feet

b. $3.14(2.5)^2(1) = 3.14(6.25)(1) =$ **19.625 cubic feet**

7. a. square inches

b. $\frac{1}{2}(36)(21) =$ **378 square inches**

Making Connections: Estimating with Pi (π) p. 211

1. (2) 25.12 feet

Estimate:
$C = \pi d$
$C \approx 3(8)$
$C \approx 24$

The closest choice is 25.12 feet

2. (3) 615.44 square inches

Estimate:
$A = \pi r^2$
$A \approx 3(14)^2$
$A \approx 3(196)$
$A \approx 588$

The closest choice is 615.44 square inches.

3. (2) 7,850 cubic meters

Estimate:
$V = \pi r^2 h$
$V \approx 3(10)^2(25)$
$V \approx 3(100)(25)$
$V \approx 7,500$

The closest choice is 7,850 cubic meters.

The Coordinate System pp. 212–213

Part A

1. A = (2,2)

2. B = (−3,0)

3. C = (−1,−1)

4. D = (2,−4)

5. E = (5,1)

6. F = (−2,4)

Part B

7–12.

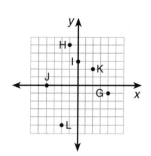

Part C

13. 6 units

14. 8 units

15. **10 units**

$$6^2 + 8^2 = c^2$$
$$\sqrt{36 + 64} = c$$
$$\sqrt{100} = c$$
$$10 = c$$

16. **(2) between 6 and 7 units**

Use points R, S, and T to form a right triangle. The legs of the triangle are SR, which equals 6, and RT, which equals 3. Use the Pythagorean theorem.

$$\sqrt{6^2 + 3^2} = c$$
$$\sqrt{36 + 9} = c$$
$$\sqrt{45} = c$$

You know the $\sqrt{45}$ is greater than 6 because $6^2 = 36$ and less than 7 because $7^2 = 49$.

17. Figure PQSR is a rectangle. You can find the distance between the points to find the measure of the sides of the rectangle. The length of the rectangle is 8 units and the width is 6 units. Use the formula for finding the area of a rectangle.

$A = lw$

$A = 8(6)$

$A = 48$ square units

If each square represents 1 square inch, the area of PQSR is 48 square inches.

Slope and Intercept pp. 214–215

Part A

1. $\frac{1}{2}$

$m = \frac{3 - 1}{3 - (-1)} = \frac{2}{4} = \frac{1}{2}$

2. $\frac{-3}{1}$ or **–3**

$m = \frac{0 - 3}{0 - (-1)} = \frac{-3}{1}$

3. $\frac{1}{3}$

$m = \frac{0 - (-2)}{3 - (-3)} = \frac{2}{6} = \frac{1}{3}$

Part B

4. y-intercept $= (0,4)$

5. x-intercept $= (1,0)$

Part C

6. y-**intercept** $= (0,2)$

$2x + 3y = 6$
$2(0) + 3y = 6$
$3y = 6$
$y = 2$

x-**intercept** $= (3,0)$

$2x + 3y = 6$
$2x + 3(0) = 6$
$2x = 6$
$x = 3$

7. y-**intercept** $= (0,-4)$

$x - y = 4$
$0 - y = 4$
$y = -4$

x-**intercept** $= (4,0)$

$x - y = 4$
$x - 0 = 4$
$x = 4$

8. y-**intercept** $= (0,3)$

$3x + y = 3$
$3(0) + y = 3$
$y = 3$

x-**intercept** $= (1,0)$

$3x + 0 = 3$
$3x = 3$
$x = 1$

9. Yes. Both the x- and y-axis share point (0,0). Imagine a line that passes through point (0,0). That line will have the same x- and y-intercept.

Unit 5 Review pp. 216–217

Part A

1. –64

2. –18

3. 10

4. 13

5. 30

6. 45

Part B

7. **(3)** $n - 15 = 85$

8. **(1)** $\frac{80}{n} + 15 = 35$

9. **(2)** $12x = 121$

10. **(3)** $3x + 18 = 9x$

Part C

11. $y = 10$

12. $a < \frac{5}{6}$

13. $x = -10$

14. $b \geq 1$

15. $z = \frac{4}{3}$ or $1\frac{1}{3}$

16. $n = 2$

Part D

17. $2a^2 - b = 15$
$2a^2 - 3 = 15$
$2a^2 = 18$
$a^2 = 9$
$a = 3$

18. $2x + y = 2$
$y = 2 - 2x$
$5x - 4y = 31$
$5x - 4(2 - 2x) = 31$
$5x - 8 + 8x = 31$
$13x = 39$
$x = 3$

19. $c - d = 4$
$2d - 2 - d = 4$
$d = 6$
$c = 2d - 2$
$c = 2(6) - 2$
$c = 12 - 2$
$c = 10$

20. $5s - 3t = 14$
$5s - 3(5s + 2) = 14$
$5s - 15s - 6 = 14$
$-10s = 20$
$s = -2$
$t = 5s + 2$
$t = 5(-2) + 2$
$t = -10 + 2$
$t = -8$

Part E

21. **56.52 inches**
$C = \pi d$
$C = \pi(18)$
$C = 3.14(18)$
$C = 56.52$

22. a. 25°
$90° - 65° = 25°$

b. 72 centimeters
$48 + 24 = 72$

23. 36 feet
$\dfrac{\text{Shadow}}{\text{Height}}$ $\quad \dfrac{42}{x} = \dfrac{7}{6}$
$\dfrac{42(6)}{7} = 36$

24. 216 cubic meters
$6^3 = (6)(6)(6) = 216$

25. a. isosceles

b. 30°
$75° + 75° + x = 180°$
$150° + x = 180°$
$x = 30°$

26. 44 square yards
$8\left(3\frac{1}{2}\right) + 8(2) = 28 + 16 = 44$
$or\ 8\left(1\frac{1}{2}\right) + 16(2) = 12 + 32 = 44$

27. 48 feet
$a^2 + b^2 = c^2$
$c^2 - a^2 = b^2$
$\sqrt{50^2 - 14^2} = b$
$\sqrt{2{,}500 - 196} = b$
$\sqrt{2{,}304} = b$
$48 = b$

28. 5 units
$a^2 + b^2 = c^2$
$\sqrt{3^2 + 4^2} = c$
$\sqrt{9 + 16} = c$
$\sqrt{25} = c$
$5 = c$

Working Together

Answers will vary.

Glossary

acute angle an angle measuring less than 90° (p. 185)

angle a figure formed when two rays meet at a single point (p. 185)

area the amount of surface of a two-dimensional figure. The area of a rectangle is length times width (p. 206).

4 ft. × 3 ft. = 12 sq. ft.

average a typical value that represents a group of values. To find the average, add the values, then divide by the number of values in the group (p. 57).

Monica scored 81, 94, and 83 on three tests.
Her average score is 86.
81 + 94 + 83 = 258
258 ÷ 3 = 86

capacity the amount of liquid (such as water) or granular substance (such as salt) that a container can hold (p. 118)

1 gallon   1 liter   1 cup

circumference the distance around a circle (p. 190)

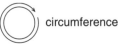
circumference

compatible numbers pairs of numbers that are easy to work with. The pairs are based on basic multiplication and division facts (p. 23).

210 and 7   12 and 600
(210 ÷ 7 = 30)   (12 × 50 = 600)

complementary angles two angles that add up to 90° (p. 188)

∠a and ∠b are complementary.

compound probability the chance of two or more events occurring (p. 149)

cube a box whose sides have equal length, width, and height (p. 208)

Each side of the cube is a square.

decimal a number written with a decimal point. Decimal values smaller than 1 are written to the right of the decimal point (p. 34).

decimal point

0.75
read "seventy-five hundredths"
This decimal is equivalent to $\frac{3}{4}$.

decimal point a point that separates the whole number from the fractional part or dollars from cents (p. 34)

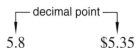

5.8 $5.35

denominator the bottom number of a fraction. It indicates the total number of parts (p. 66).

$\frac{3}{4}$ ◀— denominator (4 total parts)

diameter the distance across a circle through its center (p. 190)

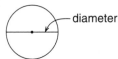

digit a symbol for a number. The digits are 0, 1, 2, 3, 4, 5, 6, 7, 8, and 9 (p. 14).

discount the amount by which a price has been reduced (p. 106)

regular price − discount = sale price

dividend the number being divided in a division problem (p. 56)

dividend
↓
$28 \div 4 = 7$

divisor the number that you are dividing by in a division problem (p. 22)

divisor
↓
$48 \div 6 = 8$

English system the system of measurement most commonly used in the United States (p. 114)

equal sign the symbol (=) that shows that two amounts have the same value (p. 43)

$12 \times 5 = 80 - 20$
↑
equal sign

equation a number sentence that contains an equal sign (p. 94)

$45 - 15 = 30 \qquad 2x + 5 = 11$

equivalent fractions fractions that are equal in value (p. 70)

$\frac{1}{2}$ and $\frac{12}{24}$ are equivalent fractions.

estimate to find an approximate answer by using rounded or compatible numbers (p. 16)

$\approx$ means "is approximately equal to"

$$295 \approx 300$$
$$\underline{\times\ 18} \approx \underline{\times\ 20}$$
$$6{,}000 \quad \text{This is an approximate answer.}$$

evaluate to solve (p. 166)

$$16x = 64$$
$$x = 4$$

exponent a number that tells how many times to multiply a number by itself (p. 162)

exponent ⌐
↓
$2^4 = 2 \times 2 \times 2 \times 2 = 16$

expression mathematical symbols used to represent a relationship between numbers (p. 58)

five less than two times a number $\quad 2x - 5$

fraction a number written as the part over the whole. A fraction bar separates the part from the whole (p. 64).

$\frac{3}{8}$ ◄—fraction bar—► $\frac{9}{2}$
proper fraction $\qquad$ improper fraction

greater than the symbol (>) that shows one amount is larger than another (p. 18)

means "is greater than" ⌐
↓ ↓
$12 > 9 \qquad 3 > -5$

hypotenuse the side across from the right angle in a right triangle; the longest side of a right triangle (p. 196)

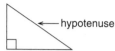

improper fraction a fraction in which the numerator is the same size or greater than the denominator (p. 68)

$\frac{15}{10}$ is an improper fraction because $15 > 10$.

inequality a number sentence describing amounts that are not equal (p. 178)

$9 > 5$ 9 is greater than 5.

$-8 < 2$ -8 is less than 2.

isosceles triangle a triangle with two equal angles and two equal sides (p. 194)

less than the symbol ($<$) that shows one amount is smaller than another (p. 18)

means "is less than"

$4 < 6$

like fractions fractions that have the same denominator (p. 74)

$\frac{3}{5}$ and $\frac{4}{5}$ are like fractions.

lowest terms not able to be reduced. A fraction is in lowest terms when the numerator and denominator cannot be simplified further (p. 70).

$\frac{5}{25}$ not in lowest terms

$\frac{1}{5}$ in lowest terms

mean the average of a set of values, found by dividing the sum of the set by the number of values in the set (p. 152)

$\$9 + \$12 + \$6 = \$27$

$\$9 \leftarrow$ mean

$3\overline{)\$27}$

number of values ┘ └ sum of the values

median the middle number of an odd set of numbers; the mean of the two middle numbers of an even set of numbers (p. 152)

Odd set: 14, 18, 21, 28, 33

median ────

Even set: $25, $28, $36, $52
Median is $32 (mean of $28 and $36).

metric system a system of measurement based on the decimal system (p. 72)

Basic units of metric measurement

for length: meter

for weight: gram

for volume: liter

mixed number a number that contains both a whole number *and* a fractional amount (p. 68)

$5\frac{1}{2}$ is a mixed number.

mode the value in a set of values that occurs most often. If each value occurs only once, there is no mode (p. 152).

negative number a number less than zero (p. 160)

−4 −3 −2 −1   0   1   2   3   4   5

negative

numerator the top number in a fraction. The numerator indicates the number of parts being discussed (p. 66).

$\frac{6}{7} \leftarrow$ numerator (6 parts out of 7 total parts)

obtuse angle an angle between 90° and 180° (p. 185)

145°

order of operations the acceptable order in which to do computation in a multistep problem (p. 58)

1. Do operations in parentheses.
2. Evaluate expressions involving powers and roots.
3. Do any multiplication or division in order, working from left to right.
4. Do any addition or subtraction in order, working from left to right.

percent a part of a whole that is divided into 100 equal parts. The symbol % means percent (p. 64).

50 percent is 50 parts out of 100 parts.
50% means 50 percent.

perimeter the distance around a geometric shape. To find the perimeter, add the lengths of all the sides of the figure (p. 169).

$6 + 3 + 4 + 8 = 21$
The perimeter is 21.

place value the value of a digit that depends on its position in the number (p. 14)

135.7   The digit 3 represents 3 tens, or 30.

probability the likelihood (or chance) of something happening or not happening (p. 112)

The probability of rolling a 4 is 1 out of 6.

proper fraction a fraction in which the numerator is smaller than the denominator (p. 68)

$\frac{9}{24}$ is a proper fraction because $9 < 24$.

proportion two equal ratios or fractions (p. 94)

$$\frac{32 \text{ miles}}{2 \text{ gallons}} = \frac{160 \text{ miles}}{10 \text{ gallons}}$$

quotient the answer in a division problem (p. 22)

radius the distance from the center of a circle to its edge (p. 190)

radius

ratio a comparison of one number to another (p. 64)

ratios: 5 to 6 $\frac{5}{6}$ 5:6

remainder the amount left over when a number doesn't divide evenly (p. 22)

4 R1 ←— 4 remainder 1
4)17

right angle the 90° angle formed when two perpendicular lines meet. A right angle is often called a corner angle (p. 185).

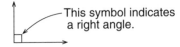

This symbol indicates a right angle.

rounding changing the value of a number slightly to make it easier to work with (p. 16)

Round 572 to the nearest 100.

$572 \approx 600$

≈ means "is approximately equal to"

266

similar figures figures that are proportional in size (p. 196)

straight angle a 180° angle; a straight line (p. 185)

supplementary angles two angles that add up to 180° (p. 188)

unlike fractions fractions that do not have the same denominator (p. 74)

$\frac{3}{4}$ and $\frac{3}{5}$ are unlike fractions.

variable a letter used to represent an unknown value in an expression or an equation (p. 94)

$c + 15 = 25$ c is the variable.

volume measurement in cubic units. The volume of a rectangular box is length times width times height (p. 208).

$V = 5 \times 3 \times 4 = 60$ cubic feet

Tool Kit

Decimal, Fraction, and Percent Equivalencies

Decimal	Fraction	Percent
.1	$\frac{1}{10}$	10%
.2	$\frac{1}{5}$	20%
.25	$\frac{1}{4}$	25%
.3	$\frac{3}{10}$	30%
.333 or .33$\frac{1}{3}$	$\frac{1}{3}$	33$\frac{1}{3}$%
.4	$\frac{2}{5}$	40%
.5	$\frac{1}{2}$	50%
.6	$\frac{3}{5}$	60%
.666 or .66$\frac{2}{3}$	$\frac{2}{3}$	66$\frac{2}{3}$%
.7	$\frac{7}{10}$	70%
.75	$\frac{3}{4}$	75%
.8	$\frac{4}{5}$	80%
.9	$\frac{9}{10}$	90%
1.0	$\frac{10}{10}$	100%

Tool Kit

Common Units of Measurement

English System

Length

1 foot (ft. or ')	=	12 inches (in. or ")
1 yard (yd.)	=	3 feet
	=	36 inches
1 mile (mi.)	=	1,760 yards
	=	5,280 feet

Weight

1 pound (lb.)	=	16 ounces (oz.)
1 ton (tn.)	=	2,000 lb.

Capacity

1 tablespoon (tbsp.)	=	3 teaspoons (tsp.)
1 cup (c.)	=	8 fluid ounces (fl. oz.)
1 pint (pt.)	=	2 cups
	=	16 fluid ounces
1 quart (qt.)	=	2 pints
	=	4 cups
	=	32 fluid ounces
1 gallon (gal.)	=	4 quarts
	=	128 fluid ounces

Metric System

Length

1 centimeter (cm)	=	10 millimeters (mm)
1 meter (m)	=	100 cm
	=	1,000 mm
1 kilometer (km)	=	1,000 m

Weight

1 gram (g)	=	1,000 milligrams (mg)
1 kilogram (kg)	=	1,000 g
1 metric ton (t)	=	1,000 kg

Capacity

1 metric tablespoon	=	3 metric teaspoons
	=	15 milliliters (ml)
1 metric cup	=	250 milliliters
1 liter	=	1,000 milliliters
	=	4 metric cups
1 kiloliter (kl)	=	1,000 liters

Comparing English and Metric Units
(English units are written first.)

Length

1 inch	=	2.54 centimeters (exactly)
1 yard	≈	0.91 meters
1 mile	≈	1.6 kilometers

Weight

1 ounce	≈	28 grams
1 pound	≈	0.45 kilogram
1 ton	≈	0.91 metric ton

Capacity

1 teaspoon	≈	4.9 milliliters
1 cup	≈	0.94 metric cup
1 fluid ounce	≈	30 milliliters
1 quart	≈	0.94 liter
1 gallon	≈	3.8 liters

Powers, Square Roots, and Order of Operations

Finding Powers

A **power** means a number multiplied by itself.

base $\longrightarrow 4^2 \longleftarrow$ exponent

To find a power, write the base the number of times indicated by the exponent and multiply across.

Examples: $4^2 = 4 \cdot 4 = 16$ $\qquad$ $6^3 = 6 \cdot 6 \cdot 6$

$36 \cdot 6 = 216$

A number raised to the second power is called a square. A whole number squared is called a **perfect square.** The table at the right shows some common perfect squares.

Finding Square Roots

The opposite of squaring a number is finding a **square root.** The symbol for this operation is $\sqrt{}$, called a **radical sign.**

Example: Find $\sqrt{64}$.

To find a square root, ask yourself, "What number multiplied by itself would result in this number?" Since $8 \cdot 8 = 64$, $\sqrt{64} = 8$.

Learning the perfect squares listed in the table on this page can help you find some square roots. For large square roots, use a calculator.

Perfect Squares
$2^2 = 4$
$3^2 = 9$
$4^2 = 16$
$5^2 = 25$
$6^2 = 36$
$7^2 = 49$
$8^2 = 64$
$9^2 = 81$
$10^2 = 100$
$11^2 = 121$
$12^2 = 144$
$13^2 = 169$
$14^2 = 196$
$15^2 = 225$
$20^2 = 400$

Order of Operations

To evaluate expressions with two or more operations, perform the operations in the following order:

Example

1 Grouping symbols $\qquad$ $\dfrac{(4 + 2)^2}{3} + 5 \cdot 4$

2 Powers and roots $\qquad$ $\dfrac{(6)^2}{3} + 5 \cdot 4$

3 Multiplication and division $\qquad$ $\dfrac{36}{3} + 5 \cdot 4$

$12 + 20$

4 Addition and subtraction $\qquad$ **32**

Within each level, work from left to right.

Tool Kit

Formulas

	Shape	Figure	Formula	Description
P E R I M E T E R		Rectangle	$P = 2l + 2w$	l = length w = width
		Square	$P = 4s$	s = side
		Polygon with n sides	$P = s_1 + s_2 + ... + s_n$	s_1, etc. = each side
		Circle	$C = \pi d$ or $C = 2\pi r$	$\pi \approx 3.14$ d = diameter r = radius
A R E A		Rectangle	$A = lw$	l = length w = width
		Square	$A = s^2$	s = side
		Parallelogram	$A = bh$	b = base h = height
		Triangle	$A = \frac{1}{2}bh$	b = base h = height
		Circle	$A = \pi r^2$	$\pi \approx 3.14$ r = radius
V O L U M E		Rectangular solid	$V = lwh$	l = length w = width h = height
		Cube	$V = s^3$	s = side
		Cylinder	$V = \pi r^2 h$	$\pi \approx 3.14$ r = radius h = height

Tool Kit

Other Formulas

	Formula	Description
 Pythagorean theorem	$a^2 + b^2 = c^2$	c = hypotenuse a and b = legs
Distance (d) between two points on a plane	$d = \sqrt{(x_2 - x_1)^2 + (y_2 - y_1)^2}$	d = distance (x_1, y_1) and (x_2, y_2) are two points on the coordinate system.
 Slope	$m = \dfrac{y_2 - y_1}{x_2 - x_1}$	m = slope (x_1, y_1) and (x_2, y_2) are two points on the coordinate system.
Mean (Average)	$\text{mean} = \dfrac{x_1 + x_2 + \ldots x_n}{n}$	$x_1, x_2, \ldots x_n$ = items n = number of items
Median		The median is the middle number for an odd set of values. For an even set of values, the median is the average of the two middle numbers.
Distance (d) as a function of rate and time	$d = rt$	d = distance, r = rate, t = time
$\$\$\$$ Cost	$c = nr$	c = cost, n = number of items, r = rate per item
Fahrenheit	$F = \frac{9}{5} C + 32$	F = Fahrenheit temperature C = Celsius temperature
Celsius	$C = \frac{5}{9} (F - 32)$	C = Celsius temperature F = Fahrenheit temperature